"Tudge is one of the most original, radical and forward looking writers on the subject of food and farming and this book is a great achievement. All politicians, C.E.O.s of food corporations, managers of supermarkets, and everyone involved in food and farming should read this book."
Satish Kumar, Editor, *Resurgence*

"Agriculture is not a business like any other, and the delusion that it is, has created bewildering problems, ranging from declining fertility of soils and increasing dependence on artificial aids to migration from country to towns and unhealthy choices of diet. Look now for original solutions from Colin Tudge."
Sir Crispin Tickell, Director of the Policy Foresight Programme at the James Martin Institute for Science and Civilization at Oxford University

"A charming polemic against 'the nonsense churned out by the modern food industry'. It whets the appetite, physical and moral."
The Guardian

"…a rallying cry to all of us who care not just about food and farming but also about Planet Earth and people everywhere."
Organic Gardening

"Thought-provoking and controversial…its central message…is hard hitting."
New Agriculturist

"…a readable, compelling book. *Feeding People is Easy* is [Tudge's] prescription for the planet."
Financial Times

"This provocative book…proposes nothing short of a new world order."
Focus

"…a closely argued examination of an all-important problem…Tudge has given us plenty to chew on."
Geographical

"…a useful and insp   ho has earned the right to thump his   a plough—more power to his elbow.
Food Magazine

I1053988

Feeding People
is
Easy

Colin Tudge

Feeding People
is
Easy

Pari
Publishing

PARI PUBLISHING

Colin Tudge cooked his first meal (an astonishingly greasy bread pudding) at the age of nine; read biological sciences at Cambridge in the early 1960s; then became a writer—mainly for *New Scientist, Farmers Weekly,* BBC Radio 3 and also for various agricultural research institutes; and so has been able to pursue his interest in food and farming and the underlying sciences and practicalities at first hand, in all the habitable continents. This book (his fourteenth, or thereabouts) is a summary of findings so far.

Pari Publishing

Via Tozzi 7, 58040 Pari, Grosseto, Italy

www.paripublishing.com

TABLE OF CONTENTS

Preface and Acknowledgements

WHY ME–AND WHY
THIS BOOK?

In the early 1970s food technologists, alias food processors, strategically poised between a growing band of nutritionists on the one hand and the newly industrial agriculturalists on the other, seemed to bestride the world. They were the scions and the heralds of science, with all its exactitude; and they were driven by the most unimpeachable principle of morality—nothing less than a desire to feed the human race. Above all, they declared in those far off days, people need protein, and the only realistic way to supply it was via "TVPs"—textured vegetable protein: a form of all-purpose chow spun from the protein of beans (mostly soya, but others too including broad beans) or of algae, fungi, or even bacteria that would be fed on surplus oil (for in those days the world's oil was perceived to be boundless). More broadly the food technologists were wont to say that farmers should think of their crops and livestock not in traditional terms—not as wheat, olives, Hereford cattle or Rhode Island chickens—but as raw materials for processing. This is where the future lay.

I like to get up early, and think. One summer dawn in 1974, gently rocking on our children's garden swing, it occurred to me that if everything the food technologists, nutritionists, and industrial agriculturalists were saying was actually true, then we would all be dead. Instead, at that time, the human population was growing faster than ever, at around two percent per year—so much so that "overpopulation" vied with the cold war in the imminence and perceived magnitude of its threat. Yet most people hadn't yet had a sniff of TVPs, and the "Green Revolution", based on new varieties of wheat and rice, had yet to make its impact. I had been educated as a biologist and was steeped in scientific orthodoxy—I had left university less than a decade before—but I realized in that moment, one of true epiphany, that all the official lore from on high, from government-approved high-tech industry, was as near as makes no difference the most absolute nonsense. Traditional farming could do with some help no doubt but it had nonetheless enabled the human race to flourish over the previous 10,000 years as no other species ever had; and traditional cooking, which the processors were so keen to replace with their own high-tech concoctions, was beyond improvement. Either the powers-that-be were seriously misguided—just plain wrong—or they were lying.

More than three decades later, genetic engineering has come of age and the powers-that-be—governments, agribusiness, scientists, and technologists, supported by economists, lawyers, and that new brand of business graduates known as MBAs—are queuing up to tell us that we will all starve unless we learn to love GMOs (genetically modified organisms, where "organisms" means crops and livestock). "GMO" has replaced "TVP" as the high-tech buzzword for salvation. It was nonsense then and it is nonsense now. Meanwhile the power within the food industry has shifted, particularly in Britain, away from the processors and towards the supermarkets; and whereas the economies of the early 1970s were varied (capitalism took many forms, while China and the USSR both had centralized, planned economies) the whole world now is hooked on or entrapped by the alleged advantages and joys of the global free market. Many have grown very rich from the high-tech, industrialized and increasingly globalized food supply chain; but they have become rich at the expense of humanity as a whole, and other species, and of this Earth, where we all live. The powers-that-be hold out the promise of boundless wealth for all if only we continue as we are—but that, in a finite world, is the greatest nonsense of all.

The message of this book, I modestly claim, is the most important that can be conceived: that we, human beings, can feed ourselves to the highest standards both of nutrition and of gastronomy; that we can do so effectively forever—for the next 10,000 years, or indeed the next million; that we can do this without cruelty to livestock and without wrecking the rest of the world and driving other species to extinction; and that if we do the job properly, we will thereby create human societies that are truly agreeable, cooperative and at peace, in which all manner of people with all kinds of beliefs and aspirations can be personally fulfilled. The approach is not to replace traditional ways of life and know-how with government-backed, industrialized high tech, buoyed by battalions of salaried experts and intellectuals, but to build upon the traditional crafts: get to know them and understand them; help them along with science of a truly appropriate kind; and practice them in societies that are intended to be agreeable. Craft is what's needed, as was always the case: but craft aided by science. The present perception of modernity—all labor performed by machines, all controlled from above—in truth belongs to the nineteenth century. The future lies with "science-assisted craft"—if, that is, we are to have a tolerable future at all.

But to achieve this we, humanity, must by-pass and generally sideline the powers-that-be. The world's most powerful governments, industries, and their attendant experts and intellectuals have their minds set on quite different goals, and are pulling in quite different directions—to a large extent completely opposite to what is really required. To be sure, the title of this book is a little hyperbolic: the human population will reach nine billion by 2050 and it won't exactly be easy to feed everybody to the highest standards. Yet it should be well within our grasp. Indeed this goal is so obviously achievable that it would surely be extraordinarily remiss not to give it a try. In this book I will explain how. But the harder task by far is to by-pass the powers-that-be. To do what needs doing we have to re-invent democracy, or rather to make it work almost for the first time in the history of civilization, for the chief rule of democracy—that we should be able to get rid of our "leaders" when they cease to function on our behalf—has gone missing. "They" do not know how to run the world, but they do know how to hang on to power. In the last chapter, I will be addressing this, too. Revolution is not required. Renaissance is what's needed—and that is very achievable.

For my own part, I am not a country boy by birth—I was born in London in World War II—but I feel I am a farmer genetically. One side of the Tudge tribe has always been farmers—some were distinguished cattle-breeders—although I belong to the side that got kicked off the land, and became miners. I first became seriously interested in food and farming in the early 1960s, when I was reading zoology at Cambridge (although wasting too much time on cooking). Later I worked for various magazines and for the BBC, who very kindly paid for me to travel and to eat in nice places, and I got to know many of the world's finest cuisines at first hand. I also worked for a time for *Farmers Weekly*, which I treated as an apprenticeship, and for *New Scientist*, and then for several years wrote scientific reports for what was then the Agriculture and Food Research Council, and subsequently for other agricultural institutions including Rothamsted (the oldest dedicated agricultural research institute in the world) and IPGRI (the International Plant Genetic Resources Institute), and spent time at research stations such as ICRISAT (the International Crops Research Institute for the Semi-Arid Tropics), in Hyderabad, India. So I have talked to farmers, agricultural scientists and policy makers in dozens of countries in every continent where farming is practiced (all except Antarctica) and attended or spoken at dozens of conferences. These included the World Food Conference in Rome in 1974—where I was shocked to find that the world's big nations did not set out to solve the problem in hand (of how to feed people) but simply to demonstrate that whatever was going wrong was not their fault, and that they should not be called upon to do more than they were doing already. In the early days, too, when my children were young and we lived in South London, I grew vegetables and kept chickens and got very involved in self-sufficiency. My first book was about farming—*The Famine Business*, published in 1977—and most of my books since have been on related issues, such as *Crop Plants for the Future* in (1988) and *Future Cook* (aka *Future Food*) in 1980, and *Neanderthals, Farmers, and Bandits* in 1998 on the origins of farming. The latest were *So Shall We Reap* in 2003 and *The Secret Life of Trees* in 2005, which discusses the growing and encouraging vogue of agroforestry.

All the time I have been arguing that the root cause of all the world's food problems has nothing to do with shortcomings of humanity, or with the innate inhospitality of planet Earth, and everything to do with policy. Largely, I have felt myself talking to the deaf. But a band of like-minded thinkers of various kinds have kept me going—such as Professor Bob Orskov who is now at the Macaulay Institute, Aberdeen, Sir Colin Spedding and Sir Crispin Tickell, and a number of farmers and fellow thinkers such as Roland Bonney and Ruth Layton of the Food Animal Initiative, Oxford, Professor Tim Lang, Geoffrey Cannon, Peter Bunyard, Satish Kumar, Robin and Binka le Breton of the Iracambi Rainforest Research Centre in the Atlantic rainforest of Brazil, Derek Cooper and Sheila Dillon of the BBC, and various luminaries at the Soil Association including Peter Melchett, Craig Sams, and Patrick Holden. Also a few years ago I married Ruth West, who is a lively thinker and critic and has kept me on my intellectual and moral toes.

But recently, certainly in Britain and the United States, I have felt the tide turning. More and more people, including some in high places, are acknowledging that the powers-that-be really have got it wrong. More and more, I have been invited to give lectures in a whole variety of venues and countries, from book festivals and farmers' meetings to university research departments, all over Britain from Shetland to Cornwall and from China to New Zealand to Brazil, and learnt a great deal from everyone I have met—it's a good way to become aware of all (or at least a great range of) the possible arguments. I have also become attached in various capacities to organizations such as the Farm Animal Welfare Trust in Oxford, the Iracambi Rainforest Research Centre, the Food Ethics Council, and the Pari Center for New Learning, established by the physicist David Peat in the village of Pari, Tuscany. All have stimulated me to develop my thinking further.

This book is a summary of thoughts so far—on what has gone wrong and why, and how to put things right. I wanted to make it as short as possible, preferably to be read in a couple of hours or less, though it has grown a bit. Maureen Doolan and Eleanor Peat at Pari have provided the opportunity, through their new company, Pari Publishing. I am very grateful to them and to Andrea Barbieri for his excellent design.

Chapter 1

GOOD FOOD FOR EVERYONE FOREVER—BUT ONLY IF WE TAKE MATTERS INTO OUR OWN HANDS

The message of this book is as positive as anyone could hope for: the future could still be glorious. We just have to do things differently.

But by "We", I mean people at large. "We"—ordinary citizens, known to the powers-that-be as "the public", to be gulled, cajoled, and ultimately blamed—have to take matters into our own hands. The powers-that-be—governments, the corporates, and the intellectuals and experts who are paid to tell them what they want to hear—are screwing things up horribly, and seem more or less bound to go on doing so. So the task is complicated because for the time being we will have to work around the people and the institutions that affect to be in charge but in truth are getting in the way.

To be sure, on the face of things, our present position could hardly be worse. At least forty years ago the "Greens" began to warn of pending disaster, and now everyone is doing it. Britain's Archbishop of Canterbury Dr Rowan Williams has talked of mass wipe-out—two billion deaths from global warming—while the President of Britain's Royal Society, Lord (Martin) Rees, suggests that our civilization has only a 50-50 chance of survival, at least in a tolerable state, beyond the present century.

The excuse from high is that the task has simply become too difficult. There are, we are told, too many of us. In addition, we are beset by forces that are beyond all control: earthquakes, tsunamis, failed monsoons. But the powers-that-be are doing the best that can be done. Their solution is for more of the same: more science, more cash, more control of natural phenomena and of people: no more "failed states"; an end to "terrorism". The world's most powerful governments, the corporates who provide the money, and the experts and intellectuals who are paid to advise them, are on the case. All will be well. Never fear. Or if it isn't, then be assured that things could have been even worse (although it isn't at all clear what could be worse than mass annihilation).

Yet the truth is almost entirely opposite. We are not suffering because there are too many of us. Ours is a restless planet and tsunamis and the rest don't help—but our world is eminently habitable nonetheless. The cause of all our troubles has almost nothing to do with the difficulties that nature presents us with. The fault lies almost entirely with policy and strategy: ideas and courses of action dreamed up by human beings and put into practice by human beings. If we, humanity, analyzed our own problems more astutely and from first principles, and if we did things differently, then even at this late hour we could create a world that was good for everyone, and for all our fellow creatures, forever. We should certainly be thinking of the next 10,000 years, or indeed of the next million. There is no good biological reason why our species should not last at least that long. That we should be seriously doubting our ability to make the world safe even for our own grandchildren is ludicrous.

I am going to focus on the world's food supply, "from farm to fork" as the cliché has it. Food is not the only thing we have to get right—of course not; for everything is connected with everything else. But food is the most pressing issue—day-by-day and even hour-by-hour: the thing we absolutely have to get right. Get farming right, and everything else we want to achieve can begin to fall into place, from the day-to-day pleasures of good eating and social living, to the grand aspirations of full and fulfilled employment, world peace, and the conservation of wildlife. Get agriculture wrong, and everything else is compromised. At present, of all human endeavours, farming is the most ill-fashioned and ill-starred of all, and the ill effects of this are indeed felt everywhere, and by everyone; and the solutions proposed and put in train by the powers-that-be are making things worse.

To begin with, though—just to get the negatives out of the way—are things really as bad as they seem? To which, regrettably, the answer is "Yes"—and probably far worse; although, through the gloom, there are a few rays of sunlight—enough to give serious grounds for hope.

The bad news

First, the world population now stands at 6.4 billion (6400 million). China and India between them account for 40 percent of the whole, with 1.3 billion in China, and 1.2 billion in India. Then there are 840 million in Africa, 710 million in Europe, 514 million in the continent of North America, 371 million in South America, and 143 million in Russia. The world total is just over twice what it was in 1950, four times greater than in 1900, six times more than in 1800, and somewhere between 200 and 600 times more than at the time of Christ. That, we hardly need telling, is a lot. But numbers continue to grow. By 2050, so the United Nations tells us, there will be nine billion of us.

Of the 6.4 billion who are with us now, so the UN calculates, roughly one million are chronically undernourished. Many simply don't have enough to eat and some suffer from specific deficiencies—up to 10 million children per year are affected by xerophthalmia (or so it is estimated) and five percent of them are blinded by it. Yet at the same time another billion suffer from the modern phenomenon of overnourishment. Obesity in many communities is the norm—and not simply, or even usually, among rich people. It's the poorer people who are fatter these days—you see fat kids in Brazil and China and among black and poor white South Africans, sometimes in the same neighborhoods where other kids are seriously underfed. It all depends what side of the burger-and-soda line you are on. Obesity *per se* is not outstandingly dangerous—but it is associated with conditions that are; such as coronary heart disease (CHD) and diabetes. CHD now accounts for an astonishing 30 percent of all deaths worldwide: 15 million per year, including 11 million in the "developing" world. Soon, according to the World Health Organization, the number of diabetics worldwide could exceed the total population of the present United States.

Then again, an estimated one billion people worldwide—one in six of all of us—now live in urban slums. Yet the cities continue to grow—with many countries actively pursuing policies of urbanization. The Chinese estimate that, within the next few decades, another half billion people will abandon the countryside for the cities—more than the total population of the newly expanded European Union; and these will be *additional* to the people who already live in China's cities. For the first time, in 2006, the number in cities worldwide equaled the number in the countryside. By 2050, on present trends, two-thirds of all humanity will live in cities: the number of city-dwellers will then exceed the total population of the present world. The population of Mexico City and a few others is already approaching 20 million. By 2050, on present trends, some will have reached 50 million, equal to the total population of present-day England. If the cities could cope, then that would be fine. City life can be highly agreeable. Already, however, it is clear that the cities cannot keep up. In the Third World in particular all the big cities are surrounded and interlaced by slums—shanties, favelas—and parts of many if not most big cities of the west are exceedingly unpleasant to live in, and often seriously dangerous (and again, the "improvements" imposed from on high over the past half century have often made things worse).

Alongside the rising numbers is rising consumption. Indeed, people in poor countries become very angry, and justifiably, when people who live in rich countries lecture them. As the Californian environmentalist Paul Ehrlich has been pointing out for more than three decades, an impeccably demographic middle-class family in Los Angeles, with Mom, Pop, and two bouncing kids, studying respectively for law and medicine and therefore of unquestionable value to humankind, consumes more than an entire Bangladeshi village. The US has only 4 percent of the world's population and yet consumes more than a fifth of all the world's energy, and the same is roughly true of resources as a whole. The British are well behind middle-class Americans in the consumption stakes, but even so it would require the resources of three planets Earth to "raise" everyone to the material standards of the average Brit. For all of us to live at the standards of rich Californians would require—well, an infinity of Earths; for there is no obvious upper limit to what people can consume if they put their minds to it.

Because of this—ever-increasing consumption, rather than mere numbers—the world's most basic resources on which all depends are seriously compromised. In 1999 the United Nations Environment Programme concluded that the lack of fresh water was second only to global warming in the league table of threats. One person in five at present has no access to water fit to drink—yet demand is liable to increase by 40 percent over the next two decades. Less than three percent of the world's water is fresh, and two-thirds of that is locked up in ice (although it is melting rapidly, though not mostly in places that are useful to humankind). Crops and livestock are especially thirsty: about 70 times more fresh water is needed to provide one person with their daily 3000 or so kcals of energy than they need for washing and so on. The next wars, so many sober analysts suggest, may well be fought over fresh water.

Land is at a premium, too. The total world area is just under 150 million square kilometers. About a third of the whole is Asia, another fifth is Africa, a sixth is North America and another eighth is South America. Europe accounts for about one-fifteenth of the whole, and Australia is just a bit smaller than Europe. About a third of all the world's land—around 50 million square kilometers—is farmed: the area cultivated or grazed has increased five-fold over the past 300 years. So now there are about 120 people for every square kilometer of farmland. Agriculture continues to spread—only about 45 million square kilometers were farmed in 1966—but we are getting to the end: most of what isn't farmed these days is desert, mountain, or city. But in 2006 the International Food Policy Research Institute (IFPRI) reported that about 40 percent of all the world's farmland is already seriously degraded. As global warming bites and sea-levels rise, the world's coastal strips, where much of the most productive farming is carried out, will disappear.

Oil, as all the world knows, is a finite resource that must run out eventually and, so many say, the "tipping point" could come in the next decade or so—or may already be passed. The tipping point, roughly speaking, is the half-way point, when it is no longer possible to go on increasing the supply: when there are no more big fields still to be found, and it becomes obvious that we are simply working through what's left. There are fossil fuel alternatives—coal, natural gas, and the huge beds of shale oil (oil locked fairly tightly into rock) that underlie North America. All surely have a huge role to play in the next few decades and centuries. But none is as cheap and convenient as liquid oil. Yet modern economies and ways of life are geared to oil—more and more and more of it—to the point of almost complete dependence. Sixty percent of Americans live in suburbs where life is impossible without a car; many drive 20 miles each way to the office, and 30 miles to take their dog to the vet. Skyscrapers that are the pride of modern cities are useless without elevators that are powered by electricity which for the most part is generated by oil. Without oil, or a very convincing substitute, most of America's most valued real estate becomes so much junk; badly located rubble.

The modern food supply chain uses a great deal of oil. The US uses 20 to 30 times more energy per head than the Third World does, and almost a fifth of the US uptake is used for food: the US expends more fuel energy on food than France uses for all purposes. Only one fifth of the US uptake is for growing crops; the rest is for packaging, transport, and storage. Of all the energy used for growing food, in the US and in the world in general, nearly a third is used to make artificial fertilizers while another third runs the tractors and combines. The rest is for irrigation (7 percent in the world as a whole, 13 percent in the US), pesticides, and so on. The industrialization of agriculture, US-style, is widely perceived as a prime indication of "progress"; already, worldwide, 95 percent of all food production depends on oil. Traditional, labor-intensive farming has become a source of shame. Of course fossil fuels aren't the only source of energy. But nothing that is on the horizon will ever be as cheap and easy as oil has been. The biggest party in the history of planet Earth is all but over. This alone is enough to suggest that if we seriously want to feed the world's people, we have to change direction.

While all this is going on, power is shifting from the West—basically the United States and Europe, with noises off from Australia—to the East: mainly to China and India. For instance—just to take one rather stark statistic: in the mid-1990s the US retailers Wal-Mart got 94 percent of its products from within the US but now, 80 percent of the suppliers on its database are Chinese. The "tiger economies" as in Japan and Malaysia have a curiously hybrid status, betwixt and between. Russia will soon be rich again from all its oil and gas—not happy, for it seems to have little gift for happiness—but rich. Right now Russia is sitting on the fence, cosy with the Chinese, anxious not to offend the US.

Britain has kept up the illusion of prosperity these past few decades by a series of con tricks. For the time being China in particular is flooding the world with goods that are extremely cheap, mainly because it has been paying its own workers extremely low wages, currently averaging $4600 per year. At the same time, Eastern Europe remains a source of cheap labor, and so too are the poor countries of Britain's former Empire, notably India and Pakistan. Brazil, too, is now exporting cheap labor. Successive British governments since the early days of Margaret Thatcher in the late

1970s, ever eager to cash in on short-term bonanzas, always strong on rhetoric but almost bereft of foresight, have run down Britain's industries partly because British workers have tended to be stroppy and hard to deal with, but also because, for the time being, it has been cheaper to buy goods and labor from abroad than at home. Britain's big and valuable coal mines were run down in the 1980s and now a great many senior politicians and civil servants would like to dispense with agriculture too, although few have the courage to say so openly.

In truth, then, this past few decades, Britain has been turning itself into a nation of hairdressers and money-changers while cashing in on its history and on the once-in-a-lifetime economic transition of China. We are temporarily richer than many other countries and can buy their goods and labor on the cheap but only because we once had an Empire, and came out on the winning side in World War II and in the Cold War—not because we are doing anything now that truly puts us out in front. This state of affairs, which depends entirely on what happened in the past, and to a significant extent on the distant past, cannot endure for much longer. China's transition, and the benison that goes with it—the flood of cheap, high-quality goods—is a one-off. It is obviously temporary: Chinese workers are already demanding better wages and soon their goods must go up in price. But Britain and other such countries will still be obliged to buy what China produces because we have given up making things for ourselves. Of course, too, the Chinese boom is not sustainable. Soon, they too will suffer from declining oil like everybody else. They are already suffering climate change, as the Gobi encroaches on Beijing, at 30-50 km a year. I have seen the maize dying on the outskirts of Beijing. The farmers know it is bound to die as the rains fail but they plant it anyway because that's what they do, and there is nothing else. When China declined in the past, the rest of the world cashed in. When it declines in the future, all the world will suffer.

The shift of power from West to East is not bad in itself. There is no good reason to suppose that the East will make a worse job of bossing the world than the West has done these past few hundred years. But the rise of the East will not be smooth and the West, particularly the US, will not go quietly. As Lord Rees points out, the chances of world conflict, as the superpowers scrabble for water, land, and particularly oil, and, more abstractly, for the status of top dog, are as great as they were at the height of the Cold War in the late 1950s, when we were all supposed to live in mortal fear of being blown to kingdom come. The US with Britain shamefully in tow seeks to distract attention from this horrendous threat and to rattle its sabres vicariously by focusing on terrorism and "nuclear proliferation": the spurious threat from countries like Iraq, Iran and North Korea. In the same way, at the end of World War II, President Truman dropped the atom bomb on the Japanese mainly to frighten the Russians. But the true nature of American angst is obvious; and the issue is not that the Chinese are bad, and the US is good, or vice versa, but that both feel the need to be on top (and where Russia will stand in this pending struggle for commercial and military dominance has yet to be seen).

On top of all this, as even US President George W Bush now seems to acknowledge, comes global warming. Predictions vary; indeed the most reliable prediction is that precise prediction is impossible. It is impossible to say whether any one place on Earth will become wetter or drier, hotter or colder. Britain is an extreme case: we could have Mediterranean climates in Scotland, with Scottish Shiraz competing with Scotch whisky. Or the Gulf Stream that flows north from the tropics and keeps Britain warm could simply cease to flow, and then Britain would be cold as Baffin Island. It seems likely though not certain that most of Africa will be arid, and possibly most of Brazil as well: the already dry forest of the Cerrado reduced to desert; the Amazon forest broken into scattered woods, as it probably was during warmer times in the past.

We can be sure only that we are in for a century or so of extremes, as the difference between the world's hot spots and cold spots increases: more violent hurricanes, more snow-bound winters, even hotter summers. It is clear, too, that ocean levels will rise as the ice-caps melt—perhaps by a metre by the end of the twenty-first century and perhaps, if the Greenland ice melts, by up to six metres in the fullness of time. A one-metre rise will seriously encroach on the world's coastal strips, where half of all human beings live, and which include much of the world's most productive farmland. One metre will be enough to wipe out quite a few small island states, and seriously to embarrass many of the world's major cities, including London, New York, Amsterdam (and of course Venice). A six-metre rise would virtually wipe out Bangladesh, with its current population of 190 million—little fewer than the United States.

But long before we face such dramas we can be sure that many of the world's crops will start to fail. All modern crops live at their physiological limits. Change the conditions ever so slightly and they are likely to give up. It isn't just the changing temperature that can affect them. It's the new combinations of conditions—warm temperatures at times of shortening day length, when they would normally expect the weather to be cold; unseasonable rainfall, too much or too little. A Canadian agriculturalist told me a decade ago that he fears for the Canadian wheat belt—and much of the world depends on the Canadian wheat crop. To be sure, it should often or even usually be possible to breed new crops to cope with novel conditions. But not if the conditions are too extreme. Even if they are not too extreme in principle, it takes at least a dozen years to produce a new variety that will yield reliably in new conditions—and then the seed has to be multiplied, and multiplied again, to provide enough to sow the world's vast fields. Crops can fail instantly. We may not have a dozen years to play with. Genetic engineering, for all the hype that has attended it, cannot and will not provide any instant fix. It is just a technique. It is strategy and foresight that count.

Yet the word seems to have got around that food is just a matter of money. Bob Geldof, rock musician turned saviour, recently opined that if the hungry people of Niger had money they could buy food. Well so they could, as things are at present. But they cannot buy food if there is none to buy. Neither can anybody else—not even Britain, for all its residual wealth. If there is food to be bought on the world market when global warming starts to bite, who will be able to out-bid the Chinese? There can be no guarantees that *anyone* will be able to eat. Common sense suggests that in times of such uncertainty it would be prudent to grow crops everywhere where they can be grown, in the hope and expectation that at least some of them will do well (although we cannot predict which ones); and it makes sense, too, to grow as many different types of crop as possible—different species, and different varieties within species—so as to spread the risk. In general, in the face of great uncertainties, it is hard to improve on common sense. In reality, present-day policies are flying absolutely in the face of common sense, as we grow more and more of the same kinds of crop in greater and greater quantities in fewer and fewer places, as if the world will go on forever as it is now. The powers-that-be, it seems, do not read the newspapers or if they do, prefer to live in their own dream-worlds, intent on the next election or "the bottom line", taking advice from people who have a vested interest in the status quo, relying on institutions that have too much inertia to change direction. I will come back to all this.

So here, in brief, is the catalogue of present-day threats. We could throw in a few more—not the least of which is or are infections: AIDS; livestock-born diseases such as BSE/ CJD and bird flu; tuberculosis— an infection that thrives on poverty, malnutrition, and over-crowding; malaria, which is ever-present; and no doubt a continuing string of novelties (of which AIDS and BSE/CJD are two recent examples) as we inevitably increase our contact with wild creatures, and as existing pathogens mutate. But the broad power struggles of the world, the diminishing resources, climate change—and the effect that all of these are having on farming and hence on our food supply— are enough to be going on with.

So where is the good news—these "rays of sunlight"?

The good news

Oddly enough, one good reason for hope lies in the matter—the very large matter—of human population.

For when the world population reaches nine billion, round about 2050, it should stabilize. Nine billion should be the greatest number there will ever be. For the first time in 10,000 years—since farming first began on a large scale—the problem of feeding everybody can be seen to be finite. We, humanity, and the world at large, will no longer be feeding a population that seems destined to grow indefinitely. We, and the world at large, can heave a mighty sigh of relief. This, at least, is what the United Nations demographers are now telling us.

Populations grow when the birth rate exceeds the death rate—so much is obvious. Even the smallest excess, even in slow-breeding creatures like elephants and humans, produces a remarkable rise in numbers in a remarkably short time. If numbers grow by a trivial one percent per year—a hundred individuals at the start of each year and a hundred-and-one at the end—then the population will double every 80 years. So it was that when human beings first began to farm on a significant scale, at the end of the last Ice Age, world numbers stood at an estimated 10 million. By the time of Christ, the world population was an estimated 100 to 300 million—a 10 to 30-fold increase, achieved over 8000 years. By 1800 AD world numbers reached 1000 million—one billion: a further three-fold increase in 1800 years. The population took another century to double again: two billion by 1900, the start of the twentieth century. The next doubling, to four billion, was achieved by the 1970s—and the United Nations held its first World Food Conference in Rome in 1974.

So between the time of Christ and the twentieth century numbers increased twenty-fold. But that is not all that happened. Particularly after the agricultural and then the industrial revolutions of the seventeenth century onwards, the percentage rate of increase, increased: that is, the time that it took the population to double was steadily reduced. The percentage rate of increase reached its maximum in the 1960s—at around two percent per year. Such a rate doubles the total in 40 years. So if numbers had continued to grow at the 1960s rate, we could have expected a population of around 8 billion by 2010, and 16 billion by 2050, and 32 billion by 2090, and 60-odd billion by about 2130, and 120 billion-plus by the end of the twenty-second century. I have met agriculturalists who put somewhat too much faith in modern technology who suggest that the world could fairly easily sustain a world population of 20 billion. I have never met any who think that 30 billion would be sustainable. Yet that is the kind of figure that would have been reached within the lifetimes of our grandchildren, if 1960s rates had continued. One hundred billion—the number projected in our great-great grandchildren's time—is beyond fantasy.

If things had gone on in that vein then, frankly, I don't know what would have been our best strategy, moral or technical. Perhaps we should never give up. St Paul stressed that hope is one of the three prime virtues (along with faith and charity). But it is hard to envisage any course of action that would have been effective, or indeed acceptable. Draconian curtailment of our reproduction, or a lemming-like procession to collapse: a foul choice.

But it seems we have been spared all that. The percentage increase of growth has steadily diminished since the 1960s and by 2050 it should, on present trends, be down to zero. If the percentage increase is zero then, by definition, numbers stabilize. It will not be that easy to feed nine billion people well and forever—the title of my book exaggerates somewhat—but it should be well within our grasp. It should be eminently do-able, even in the face of global warming and diminishing oil. Given that it is do-able, and that the alternative must imply some kind of disaster, we surely should give it a go.

There is a second serendipity. Intellectuals of many kinds have been worried about "overpopulation" at least since the start of the nineteenth century when the English cleric Thomas Malthus first suggested that while the human population was growing geometrically (by a proportionate increase each year), the food supply could increase only arithmetically (by a fixed amount each year). Sooner or later, he concluded, starvation must set in. Ebenezer Scrooge, created by Charles Dickens in 1843, spoke of "the surplus population": surplus, that is, to what was needed to keep the new factories running, and surplus to what could readily be fed (within the context of mid-nineteenth century agricultural policy). Throughout the nineteenth and twentieth centuries many have defended and even welcomed war, epidemic, natural disaster and famine on the grounds that they reduced the numbers of people who would otherwise have starved anyway, and led others to starve as well.

It turns out, though, that war, pestilence, catastrophe and famine do not contain numbers effectively. Indeed, they don't work at all. Wars and epidemics can reduce numbers dramatically—but people bounce back after them. The Black Death of fourteenth-century Europe reduced the population by a third, but it was only a blip on the demographic chart. Within a few centuries Europeans were overflowing, in particular into the Americas. A baby boom followed World War II. With a few exceptions, such as the rich, Catholic Kennedy family, it is generally the people in the poorest and most desperate societies who have the biggest families—partly because they fear, rightly, that many of their children are liable to die; partly because children represent security in old age, in societies without pension schemes; and partly because, in many poor societies, women have no status at all except as mothers—and the more they multiply the more their kudos grows.

In reality, the birth rate goes down when people feel *more* secure; when they are not hungry; when their children are not liable to die; and when women have greater opportunities for personal fulfillment, apart from increasing the family. Technology helps a great deal of course—access to reliable contraception—but it seems significant that the country that was first to reduce its birthrate to below replacement level was Catholic Italy where, one might suppose, contraception would be frowned upon. Germany's population would now be falling if the Berlin wall had not come down, and brought in the East Germans. The only modern country that seems to buck the trend is Russia, whose population is falling even though its people are neither affluent nor contented. The Russians seem to have stopped breeding out of despair; but I know of no formal studies on the reasons for their falling birthrate, and should not speculate.

Overall, the picture is encouraging: that people have fewer children when things are going well; not when things are going badly. The changes that produce such a shift—security, affluence, opportunity—are the kind that would be entirely welcome on their own account. The effect on containing population is a bonus—but an enormous one.

In short: the measures needed to contain population and to give the human species a good chance of survival into the twenty-second century and beyond are entirely benign. Cruelty and neglect are counterproductive. All this suggests, too, that when we human beings are just given a chance to live our lives as we would like, then we manage our affairs very well. It is only when people are stressed that things start to go wrong—or when they are coerced from above. The women in China who are now being ordered to have only one child are the daughters and granddaughters of women who were told, by Mao Zedong, to have as many babies as possible.

The final, huge reason for encouragement lies within the human race itself. At least within the western, Christian tradition, prophets, intellectuals of all kinds, and politicians have been queuing up to tell us what a bad lot we are. The Old Testament told us we are conceived and born in sin—a concept seized upon a few centuries after Christ by St Augustine, and developed in their different ways both by the Roman Catholics and, in the sixteenth century, by John Calvin. A key theme of the Enlightenment of eighteenth-century Europe was that human beings in a state of nature are basically savages in all the pejorative senses, all at each others' throats; and would continue to be so were it not for the soothing and smoothing hand of Civilization. Civilization in turn (it was taken to be self-evident) depended on the elite, intellectuals and born leaders, who alone could be relied upon to rescue the mob from its own venality. Disdain on the one hand; conceit on the other. This attitude persists. Politicians continue to wag their fingers at us as if were all idiots and natural-born miscreants and in the apparent belief that they really do know better.

Charles Darwin, though himself a fine and humane liberal, who for example faced up to slave-owners on their own turf in South America, managed inadvertently to add fuel to the flames. In his seminal work of 1859, *On the Origin of Species by Means of Natural Selection*, he stressed the role of competition in shaping evolution. "Survival of the fittest" is how the philosopher Herbert Spencer summarized natural selection, and Darwin himself later adopted the phrase. Various kinds of intellectuals and politicians were thus confirmed in their belief that they had risen to the top of society because they were innately superior—and that, being superior, they had a "natural" right to bash the weak. Thus disdain, conceit, and solipsism had apparently been reinforced by science—which, as the nineteenth century gave way to the twentieth, was taken to be synonymous with rationality, and therefore to be incontrovertible.

Many brave souls spoke out against all this, from Jean-Jacques Rousseau in the eighteenth century through various Romantic and Socialist movements in the nineteenth and twentieth centuries, and writers, thinkers, and artists such as John Ruskin, William Morris, Mahatma Gandhi, and Ivan Illich. Between them they have pointed out that "ordinary" human beings are actually rather brilliant. In every sphere—building, farming, fishing, milling, cooking, carpentry, metal-working, music-making, tanning, weaving—we find that the basic skills are immensely subtle, and so too is the understanding that goes with them. On the moral front, common experience tells us that people are not innately "savage". People at large simply do not need self-appointed proctors to tell them how to behave. So it was that on June 15 1944, after hiding in an attic for several years while the Nazis rampaged through Amsterdam, Anne Frank wrote in her diary: " … in spite of everything I still believe people are really good at heart". I am sure she was right: and right to keep her faith in humanity. Ordinary Germans weren't Nazis. Nazism was a flight of fancy, of politicians and intellectuals. The sin of ordinary people was to follow their leaders. Alas, it is a mortal sin. For my part, I have wandered into many a village in many a remote quarter of the world and only in one particular corner of Britain where the English are seen as the enemy have I been treated with less than courtesy. A particular village in India stands out in my memory where the headman, who was about four feet tall and lived in a house of about the same height, insisted on introducing me and my companion to all of his seemingly endless family. In remotest Turkey my daughter and I bought tea in the village café—except that we weren't allowed to buy it, because we were ushered in as guests, and plied with honey cakes for good measure.

If most people are as nice, and sensible, as I believe is demonstrably (and theoretically) the case, then it seems to follow that if only the will of the people at large could prevail, then the world should be a much better place. It boils down to democracy: a central task, indeed the *sine qua non*, is to make democracy work.

You may feel this notion is simply a flight of fancy—but there is very good reason for thinking it is the case. First, as a matter of historical interest, Immanuel Kant in the late eighteenth century predicted that any nation that was truly democratic would never initiate war. Of course it would fight in its own defence when attacked, but if the people's will truly prevailed then the people would never choose to go to war unless they had to. Modern studies suggest that Kant was absolutely right. So it was that in the 1990s Professor Randolph J Rummel of the University of Hawaii examined 353 wars between pairs of nations between 1816 and 1991 and found that no two, true democracies had ever fought each other in all that time. Democracies had fought non-democracies, and non-democracies had fought each other, but democracies never make war on other democracies. (Rummel defined a democracy as a state with universal suffrage and a free press, and a "war" as any conflict that caused more than 1000 casualties, which all seems fair enough.)

On the present issue—the world's food supply—the Nobel Prize-winning economist Amartya Sen has pointed out that famine simply does not occur in democracies. Full stop.

But we denizens of the western world should not be feeling complacent. As Thom Hartmann chillingly spells out in *What Would Jefferson Do?* (Three Rivers Press, New York, 2004) the United States and Britain have been rigorously undermining their own democracies this past 30 years, and particularly after the attack on the World Trade Center on September 11, 2001. The governments have taken more and more power to themselves, systematically undermining rights that we used to consider fundamental, including rights of assembly. Demonstrably, too, both countries miserably fail the test of democracy proposed by the Austrian-British philosopher Karl Popper. The point is not, said Popper, how a society elects its own leaders, or even whether it elects them, but whether it can get rid of them when it decides that they no longer serve its interests. At the time of writing, only small minorities actively support Bush and Brown. By contrast, tribal chiefs often seem to have autocratic powers but (so many an anthropological study has shown) it usually turns out that if the chief once loses the trust of his people, then his reign ends instantly.

I may seem to digress somewhat. But my proposals for rescuing the world include the idea that democracy is key; the necessary changes can be brought about only by democratic means. It is good to have independent and scholarly support for the idea that democracy really does work—if we can only install it in the first place.

Craft, too, is a vital and related concept. Crafts evolve: they represent the collective skills and knowledge of entire societies. They are by their nature democratic. Agriculture as practiced through all but the last few decades of the past 10,000 years has been a craft industry. Modern commercial scientists and the companies that employ them like to give the impression and perhaps believe that the world's farming was in a dreadful mess until they came along, beginning in the late nineteenth century but particularly in the mid twentieth, and rescued us.

Again, the truth is quite opposite. Modern agricultural science has succeeded insofar as it seems to have done only because it had thousands of years of traditional craft to work with. By the time modern science came on the scene, wild plants and animals had already been tamed and re-fashioned and the fields made ready—not by university departments and teams of corporate scientists but by ordinary farmers, tackling life's problems; and before them by hunters and gatherers, who had worked out what is edible and compliant and what is downright dangerous. At the other end of the food chain the world's great chefs rightly grow rich from their artistry—but the best of them emphasise that they build on tradition: that their core skills and their finest recipes were devised over centuries by "ordinary" cooks in a hundred million kitchens, in Italy and Provence and Turkey and China and all the rest. The Medicis and the other great families of Renaissance Italy built Florence and Siena and all the other great and enviable cities amidst the fields of a peasant economy—but it was only because they left the peasants alone to get on with their work that the glorious civilization that we have inherited was able to arise at all.

In truth, too, Darwin's *Origin of Species* when properly understood does not predict that human beings should fight, all against all, unless otherwise restrained. All creatures depend on other creatures for their survival. All must cooperate as much as they compete. Social animals especially must emphasise cooperativeness—and human beings are more social than most. Human evolution, we should properly predict, is *bound* to produce sociality. Evolutionary theory predicts that most human beings, in any one society, are bound to be nice, because no society can hold together except by sociality and cooperativeness. Unfortunately, theory also predicts that nice people are bound to be dominated by nasty people: the minority who are more aggressive, and want to be out in front. There is the rub, the central dilemma of humankind. More of this later.

Meantime I want only to note this second outstanding reason for hope: that we can, in the end, trust ourselves. History, common experience, and evolutionary theory combine to tell us that we, people at large, are immensely ingenious and morally sound. We can take matters into our own hands. It is safe to do so.

So what, in practice, do we need to do? First, we need to get the moral philosophy right—to work out what we should be trying to do and why; and then we need to address the practicalities. I will look at the underlying philosophy in the next chapter. The practicalities occupy the rest of the book.

Chapter 2

WHY SHOULD WE
GIVE A DAMN?

The task before us—where "us" means humanity—is to create a world that is good for 6.4 billion people now, and for 9 billion people by 2050, and for the estimated five to eight million other species with whom we share this Earth; and can go on catering for everyone and for other creatures forever—at least for the next 10,000 years, and preferably for the next million and beyond. That might not be possible, but it looks possible enough to be worth a try. The prize would be very great, while the price of failure exceeds our imagining. Technically, the task should be within our compass. So why don't we set about it?

One very strong reason is that a lot of people don't frame the problem in this way. Many don't agree that we, humanity, should be trying to cater for everybody, and for all other creatures. In practice, we don't need a majority to agree on what should be done, in order to make the necessary changes. Critical mass is what's needed. I know many who do agree that the world's task is to cater for everybody and there are surely enough in the world to do what needs doing—if only their thoughts and efforts could be coordinated.

But how could anyone disagree—that it is humanity's task to care for humanity, and the rest of the world where we all live? This is worth asking. It is useful to anticipate opposition.

How could anyone not give a damn?

Some—only a few, perhaps, but they are out there—just don't give a damn. "Not my problem, Mate," I have heard people say. Others do give a damn but seem too modest: "Too much for my old head"—or as my maternal grandfather was wont to say, "The government has got something up its sleeve, Sonny, don't you worry about that!" Others agree with Voltaire, as in *Candide*: that every man should cultivate his own garden (and indeed, if more people did follow Voltaire's advice, including Bush and Brown, the world would surely be a safer place). I was brought up in a Christian tradition and Christians take it for granted that where there is a moral wrong, then they should take it upon themselves to put it right. But Christ was more interventionist than most. Many believe that it is not morally proper to intervene unless specifically invited to do so. Some point out—with some justification—that you can never fully predict the effects of your actions, particularly when it comes to dealing with other human beings, so it is better not to interfere; just go with the flow. As the proverb has it, "The road to Hell is paved with good intentions". Everybody hates a do-gooder. So there are all kinds of reasons for standing back and although some of them seem simply to be rooted in indifference and selfishness, others do have at least some venial caution behind them.

Others take a hard line. They positively, and sometimes adamantly, disagree that individual people *should* concern themselves with society as a whole, or with humanity as a whole. Some simply do not accept that cooperation is a positive force for good, or that society is more than a collection of solipsists, or that each individual should, in practice, make some concession to the whole. Compassion, they are inclined to argue, is namby-pamby, and ultimately destructive. Some appeal to Darwin (although Darwin would surely be turning in his grave): insisting that it is competition that leads to improvement, and that it is destructive to be generous to the losers, for that fosters weakness that in the end must destroy us all. Friedrich Nietzsche put a more high-sounding spin on this. He argued that God is dead, and therefore that the only judge of humankind is humankind. Among human beings, some prove to be superior—at least as judged by their own standards for there are, he insisted, no other standards by which to judge: and such people— *Ubermenschen*: "overmen"; supermen—not only have a right, but a positive duty, to develop their own talents to the full, even though this may require them to ride rough-shod over the rest. Nietzsche did not invent the *Ubermensch* mentality, although I believe that he did coin the word. It is present in European imperialism, as we casually plundered the rest of the world—not in the Nietzschean belief that God was dead but with the conviction that because we were obviously superior to Africans and Asians and native Americans and Australians then we must be closer to God's image, so He would obviously be on our side. In modern times the US "neoconservatives" who now rule the White House, openly parade their intention to dominate, forever, in the guise of the world's only superpower; and those who get in the way are pushed aside. I know many Americans who are ashamed of this attitude bur have also met many others who simply accept that the neocon attitude is the way of the world.

In short: there are many, including some of the world's most powerful people, who for a whole raft of reasons do not agree that the task before humanity is to take care of all humanity, and of our fellow creatures. It isn't just that the scales have not yet fallen from their eyes. Some respectable philosophers have offered coherent reasons why universal compassion is positively offensive. They perceive it as a philosophy of weakness: ultimately destructive. Neither do you need to be an ardent atheist, as Nietzsche was, to espouse such views. Bush, Blair and Brown declare themselves to be Christians.

How do we know what's good?

But since there is more than one moral position to be taken, doesn't this prove the relativist argument, that there are no moral absolutes? I would like to argue this further—nothing is more important than the search for moral absolutes—but since this is not a book of moral philosophy I will make just a few quick points. First, despite the best efforts of some very fine philosophers such as Immanuel Kant, we cannot hope to discover or to define moral absolutes as if they were scientific laws. Kant's own suggestions (his "categorical imperatives") in the end remain arbitrary. Many have posited that moral absolutes are what God says they are, by definition. But God's will is open to interpretation. Blair and Bush have sought divine inspiration for all that they do, or so they tell us.

Many realized by the end of the eighteenth century, hard on Kant's heels, that the search for moral absolutes was ultimately forlorn. Among them was England's Jeremy Bentham, who founded the modern school of "utilitarianism", sometimes known more generally as "consequentialism". In utilitarian philosophy, actions are judged not by some absolute (but inevitably arbitrary) moral criterion, but according to their outcome. Human happiness was and is the goal, said the humanist Bentham: and the ideal is to achieve "the greatest happiness of the greatest number".

This seems straightforward enough, and utilitarianism has carried great weight. Yet there are serious problems with it. One is to define happiness, and to measure it. A modern solution to this problem is simply to equate happiness with wealth, on the grounds that it is better to be richer than poorer. So we find that ethical decisions are increasingly made on grounds of cost-effectiveness, or indeed of expediency. If such-and-such a course is more profitable, or easier, then it must be right. Consumerism is rooted in this philosophy: it is taken as read that people buy only those things that make them happy, and since happiness is the ultimate good, then whatever they are prepared to pay for must be good too. I have even heard human cloning justified in such terms. Guns, too: if guns make people happy, so the US gun lobby says, people should have guns; and if people make money out of making guns, then that is justification too. Yet, we might suggest, there are good and bad reasons for being happy. No one to my knowledge, at least in public, has defended child pornography on the grounds that it makes some people happy, or that it is profitable. You might say that utilitarianism covers this case since child pornographers are in the minority, so it isn't the greatest *number* who are made happy. But numbers don't carry the case either. If twelve Nazis beat up one gypsy, then twelve people are made happy and only one is miserable. In this instance, happiness far outweighs misery, at least if we simply count heads. Yet we feel that such beatings are foul, and that child pornography is foul. They are *wrong*. The true roots of morality, we might reasonably suppose, are to be found not in the happiness or the misery of individuals, but in the feelings that lead us to judge that some actions are acceptable, and some are not.

This leads us to yet another eighteenth-century philosopher: David Hume. In the end, said Hume, morality can and must be guided by "passion": emotional response; attitude. David Hume was at the hub of the Scottish Enlightenment—the ultimate rationalist—and yet, as a truly accomplished rationalist, he acknowledged the limits of rationalism. In the end, morality is a matter of feeling, and you have to go with your feelings. In the end we just have to state our moral positions in the way that Thomas Jefferson and his fellow authors did in 1776 when they framed the American Declaration of Independence: "We hold these truths to be self-evident…". If only the US had stayed true to Jefferson the world would be a very different and better place. I do not presume to improve on Hume or on Jefferson. I will go with my feelings, as Hume recommended, and state, merely, that I take my conclusions in this matter to be self-evident, as Jefferson and his colleagues were content to do. It is right to cater for the whole human race. It is wrong to write people off, no matter how expedient this may be.

Two last words, however. First, if Hume is right—that morality is inevitably rooted in feeling—then it seems to me that if we truly want to refine our moral positions, then we should set out deliberately to cultivate our emotional responses. This takes us into the realms of "virtue ethics": ethics based not on absolutes or on cost-effectiveness—who gains, who loses, insofar as these things can be measured—but on attitudes. The moralists who over the past few thousand years have focused on essential attitudes include philosophers such as Aristotle and Lao-tzu. But in the main, it has not been the philosophers, but the prophets, who have stressed the need to approach life's moral problems in the right frame of mind. Atheists and other detractors are wont to argue that the prophets of different religions all pull in different directions—that what is acceptable to a Christian is taboo to a Muslim, and so on. But, invariably, the differences the detractors point to are merely matters of customs and manners—who eats pork or beef and who doesn't. At the deepest level, where morality truly begins, all the great religions are in agreement. The Ten Commandments of Moses urge us to *love* God, and *honour* our parents. Buddha spoke of compassion. Christ urged universal love. Mohammed spoke of justice and generosity. Hindus tend to be eclectic, and the nineteenth-century Hindu mystic Ramakrishna summarized them all when he suggested that the moral position of all the great religions and their prophets can be encapsulated in three irreducibly simple phrases: personal humility; respect for fellow, sentient creatures; and reverence for God (although atheists may prefer to substitute "nature" for "God").

That's it. It is remarkable how many of today's most Gordian issues, even of the most technical and modern kind, seem to unravel when exposed to such simple, ancient wisdom. Still, I do not want to argue that what the world's great prophets have agreed upon is *ipso facto* correct and incontrovertible. But I do like the idea that what they have said, to a significant extent summarizes the deep morality of all humankind. The prophets are expressing what most of us feel deep down really is *right*; and they lived in ways that most of us, deep down, would like to be capable of. Of course (the detractors will say) there have been many prophets, and they did not all agree with the founders of the world's principal religions. Indeed. But I suggest that natural selection applies. The prophets who spoke of compassion—humility, respect, and reverence—are the ones we remember precisely because what they say chimes so precisely with our deepest convictions. They express not simply what we believe, but what we are. The prophets with different teachings have fallen by the wayside. The fans of Nietzsche are in the minority.

New light is thrown on all these discussions by game theory.

Game theory, and why nasty people are in charge

Ethics—morality—is in the end a matter of interaction: how creatures that are capable of making a choice, meaning human beings, treat third parties of their own species, or other species, or indeed the world in general. All discussion of any kind of interaction—war, commerce, ecology, evolution, or human societies in general—can be assessed, and indeed quantified, in terms of game theory. We can show by means of mathematical models, obligingly run by computers with standard software, what kind of behavior is likely to bring greatest benefit, and to whom; and since the way we behave depends to a significant extent on our underlying beliefs and attitudes, game theory offers at last some insight into different moral philosophies.

One of the simplest of all game theory models is hawks versus doves. Hawks may be seen either as gangsters who simply don't give a damn, or as Nietzscheans who present coherent and high-sounding reasons for bossing everybody else around; and doves are people of the kind the prophets would approve of—humble souls, considerate of each others' feelings, anxious where possible to cooperate.

Game theory shows that societies composed entirely of hawks, collapse. The hawks spend all their time fighting for top-dog status, and beat each other up. But Bentham's ideal—the greatest happiness of the greatest number—is achieved by all-dove societies, who waste no time at all in fighting and who, by cooperating, achieve far more for their common benefit than any could achieve alone. In practice, the all-dove society cannot be achieved except by adopting dovish attitudes.

But although the all-dove society is by far the best, the theory shows that it is not stable. It is bound to be invaded by hawks, for whom the doves provide easy pickings. The hawks can take what they like in an all-dove society, without fear of reprisal. The hawks may either invade from outside—some rival tribe or nation, in human terms—or arise from within, as some of the erstwhile doves discover that hawkishness offers an easy option.

The hawks, then, being very successful, soon increase. But the theory then takes another turn. For as the number of hawks increases, so they start to get in each other's way. A hawk swaggers into what he takes to be a party of doves, demanding booty, and finds that some other hawk has already beaten him to it. So he gets in a fight. Thus (the theory has it) the number of hawks in any given society is self-limiting. When the proportion of hawks reaches, say, 20 percent, it no longer pays to be a hawk. Hawkishness lands you in too many fights. When surrounded by too many hawks it is better to be a dove—even a reluctant dove: go with the flow, keep your head down, and stay out of trouble.

I find this simplest of models most illuminating. It predicts that if societies are left to themselves they finish up not with all-hawks, or with all-doves, but with a mixture of doves and hawks. Furthermore, the doves will greatly outnumber the hawks. The theory predicts, in short, that most individuals in any society that is left to itself are bound to be nice: cooperative and pacific. But the hawks will dominate even though they are the minority, precisely because they are hawks; and the doves don't want to fight back.

Doesn't this precisely describe what happens in human societies? Most people are nice, which is what I have been arguing all along. But in any one society the nice majority is bound to be ruled by people who simply want to rule. So it is that nice people usually find themselves dominated by nasty people. The Enlightenment notion that human beings are basically bad and need to be bossed by firm leaders is the precise opposite of the truth. Most people are basically nice while their leaders very often turn out to be gangsters. It seems to follow that if we, humanity, could only create true democracies— societies that truly reflected the will of the people—then the result should be most agreeable, because most people are doves, and doves are nice. But it is extremely difficult to establish true democracies because doves are inevitably dominated by hawks. We finish up with the kind of "democracy" that we see in modern-day Britain or the US, where the masses of doves are allowed, at long intervals, to choose their leaders from a shortlist of hawks, each of whom they may find equally unsavory. How can we create democracy, then? How can doves create societies dominated by doves when doves have no taste for domination? That is the central paradox and dilemma of humankind. I will address it in the last chapter.

Meanwhile, in lieu of true democracy, the powers-that-be are convinced that it is their right and destiny to rule, and the intellectual elite whom they employ are convinced that they alone know how to do things. In truth, the leaders are liable to be gangsters, while their compliant intellectuals to a significant extent emerge as *idiots savants*. The real genius, moral and practical, lies with humanity at large. The future lies with what that other great Enlightenment moralist, Adam Smith, called "human sympathy"; and it lies, as philosophers from John Ruskin to Ivan Illich have argued this past few hundred years, with craft: the skills that have evolved among humanity at large. Since food is the thing we absolutely have to get right, the most important crafts are those of farming and cooking.

One final caveat, however. I know that many will be gnashing their teeth at this point, if they can bear to have read this far; and quite rightly. For I have met many people in high places who definitely are not gangsters, and agree absolutely that it is the proper and principal task of humanity to look after humanity. Some of them are Christians, some are Muslims, some Jews, some atheists. No matter: the common thread of human sympathy runs through all of them. But they disagree ardently with my general thesis—that the world needs re-thinking from first principles, and that humanity should put its trust in humanity at large. To be sure the present world is flawed, these critics say, but there is nothing fundamentally wrong with present-day technologies, or political or commercial institutions. Specifically, modern farming in its industrialized and high-tech form is good and necessary. The corporate food processors and supermarkets can feed the world much better than anybody else could do. All the infrastructure of modern technology and commerce are therefore necessary too—the science, the economic theory, the ever-tighter organization. The status quo doesn't need re-thinking. It just needs to be given a chance. As the *Economist* magazine declared on a recent cover (November 5th–11th 2005): "Tired of globalization but in need of much more of it". We are in a period of transition. Just have patience.

These caveats are serious. I believe that some people who disagree with me are demonstrably wicked but I also know that some at least of my critics are hugely intelligent and at least as well-intentioned as I can claim to be. So I want to take this last class of criticisms seriously. But I will do this in later chapters. First I want to present my own thesis: that the task is to make a world that is good for everyone forever; that to do this we need to think again from first principles; and that we must focus, above all, on the food supply chain—which means on farming and cooking.

Chapter 3

GREAT FOOD AND ENLIGHTENED AGRICULTURE: THE FUTURE BELONGS TO THE GOURMET

The task before us is to provide good food for everyone, forever; at the same time to create agreeable ways of life for farmers, and for everyone else involved in the food chain—and indeed for all humanity; to do this without cruelty to livestock; and to ensure that the world as a whole remains beautiful and secure, and that as many as possible of all the other species with whom we share this planet continue to thrive and to evolve. Farming is the key to all this—or at least it is the thing we really have to get right. It is the source of the thing that we need in greatest quantities, and without interruption; and it is the principal interface between humanity and the fabric of the Earth itself. The kind of farming that would do all that is necessary I call "Enlightened Agriculture".

"Enlightened" is a high-fallutin' term, with overtones both of the eighteenth-century Enlightenment of Europe, with its emphasis on reason, and on the Buddhists' spiritual path. Yet conceptually it is irreducibly simple, and technically it should be straightforward: common sense and craft within a framework of sound biology. Hence my title: "Feeding people is easy".

"Sound biology" means to acknowledge what ought to be most obvious: that human beings are flesh and blood creatures—that we are indeed animals, with the same fundamental needs as other animals; and that the world in which we live is our habitat. So it is an exercise in physiology, for we have to have some feel for what our own bodies need; and it is an exercise in psychology—for we need to respect our own desires; and in ecology—we have to know the limits of the world itself, what it can do and what it cannot. At bottom we need primarily to regard ourselves as we really are—as a biological species with in-built physical needs like any other. Like any other species, too, we must avoid fouling our own nest.

All this should be too obvious to be worth saying yet it is totally at odds—*totally* at odds—with the thinking that has underpinned agricultural strategy in the western world over the past few centuries and in particular over the past few decades. It seems as if people in high places—including many "leading" scientists, to their shame—believe the Platonic myth that human beings dwell primarily in an ethereal, spiritual-cerebral world, or should aspire to do so; and the Baconian myth that "Man" can "conquer" nature and make the world do anything we want it to, for our own comfort. As if as the *coup de grâce,* modern policy-makers have now fallen hook, line, and sinker for the myth of monetization—the belief that if human beings simply set out to make as much money as possible, in all the spheres in which they operate, then somehow or other everything will turn out all right. Driven by their myths, the powers-that-be—scientists, economists, politicians—contrive to flout the bedrock rules of biology if indeed they recognize their existence at all. For good measure they commit the cardinal sin of philosophy, which is to imagine that omniscience and hence total control are within human grasp—that they therefore can do anything, and dig themselves (if not necessarily the rest of us) out of any hole. The arrogance is stupendous.

In truth, agriculture has very rarely been designed expressly to feed people, and to my knowledge—apart from the modern organic movement—it has never expressly acknowledged the need to work within the bounds of sound biology, although this has sometimes been implicit. Most agriculture in the history of the world has been designed primarily to match and support the prevailing economic and political structure and beliefs of the day. This is equally evident in the open-field strips of Feudal Europe, or in England's great neo-Feudal proto-capitalist estates of the eighteenth century, or America's prairie homesteads, based on Thomas Jefferson's vision of the United States as "a nation of small farmers", or in Stalin's collectivist farms of the 1930s and beyond.

But never in all history have the powers-that-be had the wherewithal to operate on the global scale as they do now. Never have they been able, as now, to take the whole of world farming by the scruff of its neck and ram it so procrusteanly into a structure and a philosophy that are so alien to its purpose, and so at odds with the needs of humanity and the biological and physical constraints of the world. The powers-that-be behave as if they were playing a game—which indeed they are: a game of money and power. They are forever lecturing protestors like me about the need to be "realistic"; but the only reality they recognize is the political-economic, commercial-military power game that they happen to be engaged in, and which makes them rich. They have no feel at all for the physical realities of the world itself, and the creatures within it, and for the ways in which farming has, in reality, been practiced this past ten thousand years, and by whom. They have a great deal of "data", which they collect and publish selectively, and manipulate with the aid of lawyers and other rhetoricians this way and that, largely for our bamboozlement, but that is not the same thing at all. The world is suffering, possibly terminally, from a huge irony: that the powers-that-be live in a fantasy world of their own devising, blind to every observation that is in any way inconvenient, yet they believe that they really do know what they are doing, and that they alone are the realists. We are dying of their illusions.

We (humanity) must now take matters into our own hands—and, I believe, it is well within our power to do so. The world's food supply chain could supply the thing that has been lacking this past few thousand years—the vehicle for true democracy. But I will come to that in the last chapter. For now I want to look at Enlightened Agriculture itself: what it looks like; what it is. We might begin by looking at food—the thing that agriculture should be trying to produce. What, in fact, do human beings need?

A lightning course in nutrition

Living bodies are complicated and it takes energy to keep them together. Food provides us with the raw materials from which to construct our own flesh and with the necessary energy, expressed as "calories" (or more usefully as "kilocalories" or kcals: each kcal is a thousand calories). It also, vitally, supplies a miscellany of bits and pieces which, broadly speaking, oil the works (for example acting as intermediaries in various metabolic pathways). The components of food that meet all these requirements are roughly classed as carbohydrates, fats, and proteins, which can be called "macronutrients"; plus a very mixed bag of minerals, vitamins, and other recondite organic molecules, which are known collectively as "micronutrients".

Carbohydrates, in rock-bottom terms, are composed exclusively of just three chemical elements: carbon, hydrogen, and oxygen. In their simplest form carbohydrates manifest as sugars. Sugars joined together form "polysaccharides" which take many forms. Very common polysaccharides in nature include starch, which most plants create as a personal food-store, especially in seeds and tubers; glycogen, which animals create as a short-term source of energy, mostly stored in the muscles; and cellulose and various hemicelluloses, which form the tough but flexible cell walls of plants.

Broadly speaking, human beings need between 1500 and 4000 kcals per day depending on whether they are children or adults, growing or not growing, men or women, lactating or pregnant or neither, sedentary or sweat-of-the-brow laborers, and also on whether they naturally metabolise rapidly or less rapidly. Considering the enormous range of human sizes, shapes, and conditions, the total range of energy we require is surprisingly small—largely because big people need less energy weight-for-weight than small people, and so tend to be more economical. (This is for simple reasons of physics. Warm-blooded creatures like us use most of our energy creating body heat. Little bodies cool quicker than big bodies, whether the body in question is a human, an elephant, or a cup of coffee. So big bodies, weight for weight, are easier to keep warm and so are more economical). In most human diets (with a very few exceptions as in the traditional Inuit), carbohydrates are the chief source of energy. Whether in the form of simple sugars or of polysaccharides, carbohydrates provide roughly 420 kcals per 100 grams; so an average person (if there is such a thing!) could get all his/her daily energy needs from about 500-700 grams of carbohydrates, which is about a pound to a pound-and-a-half.

The chief carbohydrate by far in a traditional diet is starch—found mainly in the seeds and tubers of plants. Plants, then, in traditional diets, are our prime source of energy. However, human beings cannot digest cellulose (or at least, only to a very limited extent with the aid of bacteria in the hind gut); so cellulose and the hemicelluloses do not provide us with energy and old-style nutritionists were apt simply to dismiss them as "roughage". But, roughly beginning in the early 1970s, nutritionists have come to appreciate that so-called "roughage" is an extremely important component of diet even though it provides virtually no calories. In line with its new status, roughage has been re-named "dietary fibre". Carbohydrate in pure form, as in pure sugars or starch, is said to be "refined". Carbohydrate served up as it was in the original plant, with all the original fibre present, as in wholemeal bread, is said to be "unrefined".

Since human beings do not digest fibre to any appreciable extent, it increases the bulk of food without increasing the calories; and it is thereby said to "dilute" the overall energy content. Since there is a limit to the amount that people can eat—or indeed are prepared to eat—fibre thus helps to limit total intake. Indeed it is the greatest "slimming" food of all. Indeed, some people in poor countries sometimes find that their diet is too fibrous. Women in many African villagers live almost exclusively on porridge made from maize or sorghum and may find it almost impossible to consume enough calories in a day to sustain a pregnancy or a lactation (which requires even more energy than pregnancy). Their children, too, newly weaned, may find that the calories in the local porridge are just too dilute. People on such diets need more concentrated food, rather than less. In this, poor people on diets that are very high in unrefined carbohydrates, and western people who ply themselves with sugars and fats, are mirror images of each other: the poor could do with less dilution, the rich with more.

The word "dilution" could cause confusion since in most contexts "diluting" means adding water. Indeed, water also adds bulk to a diet without increasing the energy content, but its overall effect on intake seems to be ambiguous. For instance, it is probably harder to slurp a lot of watery soup than to wolf a smaller amount of more concentrated stew. But sweet drinks are another matter. It is possible to drink gallons of what Americans call soda and the British traditionally called pop, without suppressing appetite at all, and thus take in several thousand calories a day in passing—literally. These days, many people do this. Soda is comfort food, and habit-forming, not to say addictive. It takes away some of the pain of the big hot noisy city, and indeed of the air-conditioned office. Grossly obese people worldwide are typically soda-swiggers. Thus, white sugar provides 394 kcals per 100 grams; wholemeal bread provides 216; and potatoes, traditionally the slimmer's great no-no, provide a mere 80 kcals per 100 grams. However, if you fry potatoes to make what the British call chips and the Americans call fries, their energy content triples—to 250 kcals per 100 grams. Coca-cola and Pepsi provide about 175 kcals per 100 ml (which is roughly 100 grams). You can drink a liter in a day without suppressing appetite: that's an additional 1750 kcals before you even start eating: adding more than 50 percent to your required daily intake without you even noticing.

Fats too, like carbohydrates, are composed entirely of carbon, hydrogen, and oxygen. But again, these three simple elements can be combined in a virtual infinity of ways and fats too are immensely variable. Lard and suet are "hard" fats. Chicken fat is softer. The fats of fish and many seeds are so soft that they manifest as oils at room temperatures. Waxes are fats. Cholesterol is a peculiar form of fat. Chemically speaking, some fats are "long-chain", meaning they have very big molecules; and some are short-chain, with smaller molecules. "Saturated fats" contain as much hydrogen as it is chemically possible for them to contain, while "unsaturated fats" contain less hydrogen than is theoretically possible. Broadly speaking, the fats of land animals (notably mammals) tend to be hard; and hard fats tend to be highly saturated. The oils of plants and fish tend (on the whole) to be highly unsaturated. Those that are particularly unsaturated are said to be "polyunsaturated".

With the possible exception of some waxes which humans find indigestible, all fats can be "burnt" in the body as a potent source of energy. Indeed, weight for weight, fat provides twice as many calories—about 900 kcalories per 100 gram—as pure carbohydrate does. In practice, fat tends to be diluted somewhat by water, so that butter and margarine provide about 740 kcals per 100 grams—still highly calorific. Animals including humans store surplus energy in the form of fat precisely because, weight for weight, it does provide so much energy. If human beings stored energy in the form of starch, like potatoes do, we would need to carry much more of it, and would be even more rotund. Westerners have high meat diets, which almost inevitably means that we have high fat diets. So it is that while traditional rural Chinese got only about 10 percent of their calories from fat, Westerners commonly get 40 percent-plus from fat—largely animal fat. Ten percent is perhaps too low, but 40 percent is almost certainly too high. Very high fat diets evidently predispose not only to obesity, but also to various cancers (including breast cancer) and coronary heart disease. The ideal is probably between 20 and 30 percent fat (although such "ideals" are very hard to establish).

Fats, however, even more than carbohydrates, are a very important component of body structure. Every cell membrane is composed partly of protein and partly of "lipid" (which is a fancy term for fat). Brains and nerves in general are particularly rich in structural fats. However, the fats needed to build cell membranes in general and brains and nerves in particular are of a special kind, loosely classed as "essential fats". Broadly speaking, these essential fats are of the polyunsaturated kind. Both the leaves and seeds of plants provide essential oils—and leaves and seeds provide different kinds, which are not interchangeable, so we need both leaves and seeds. Fish-oils tend to be especially rich in essential fats.

Saturated fats (and much of the unsaturated) on the whole are not "essential". They serve us merely as a source of energy. But in recent years nutritionists have emphasized our need for essential fats more and more. In particular, they have tended to recommend a high intake of fish, particularly of oily fish, such as mackerel. This is sound nutritional advice no doubt, but in reality, worldwide, fish is still only a minority food; yet already, as all the world knows, many if not most of the world's major fisheries are on their beam ends and some species that used to be taken for granted are now rare in some of their traditional grounds or even extinct. North Atlantic cod is a particularly shocking example. Fish farming can add prodigiously to the supply (although some fish including cod are very difficult to farm for a variety of reasons) but raises huge problems of pollution (though these could be solved if we didn't try to do everything on the cheap). In general, though, the advice to eat more and more fish really is "unrealistic" for humanity at large. Besides, I once shared a conference platform with a nutritionist who warned the audience that their children were bound to grow up with defective brains unless they ate fish, fish, and more fish. But two of the principal speakers at the conference, who were both from inland village India and both exceedingly intelligent and eminently sane, pointed out that neither of them had encountered any fish as a thing to eat until they came, as adults, to Europe. They had got all the essential fats they needed from plants.

Proteins are chemically more complicated than carbohydrates or fats. They too are composed primarily of carbon, hydrogen, and oxygen: but they also contain nitrogen and (generally) small amounts of sulphur. In reality, like polysaccharides, proteins are not so much molecules as "macromolecules". Polysaccharides are composed of long chains of sugars, and proteins are composed of (very) long chains of amino acids. In practice, most animal proteins contain various permutations drawn from a basic library of about 20 amino acids. About eight of these, such as lysine, methionine, and tryptophan, are said to be "essential", while the rest are "non-essential". In practice, however, all of the amino acids are essential. It's just that the body cannot synthesise the ones that are called "essential" for itself, so they have to be supplied ready-made in the food. But the body is able to make the so-called "non-essential" amino acids for itself by converting one or other of the other amino acids. In traditional nutritional parlance, proteins that contain all the essential amino acids roughly in the proportion that the human body requires were called "first class"; and those that were (relatively) deficient in one or more of the essential amino acids were "second class". Broadly speaking, animal proteins tend to be first class, while plant proteins are often somewhat deficient (relatively) in one or other amino acid and so were considered to be second class.

Proteins are the main stuff (apart from water) of which flesh is made. The muscle—or at least the bit that does the contracting—is protein. Cell membranes in general are composed of protein plus fat. The hemoglobin in red blood cells is a protein. Antibodies are proteins. Some hormones are proteins. All enzymes are proteins—enzymes being the catalysts that enable the various chemical reactions to take place in the body; they are the drivers of metabolism. Clearly, if there can be degrees of essentialness, then proteins are among the most essential of all.

But there has been a revolution in thinking about proteins over the past few decades just as there has been in attitudes to dietary fibre, and to essential fats. For when I was at school and university during the 1950s and early '60s, we were told both as growing lads and as budding biologists that human beings needed to eat protein in relatively vast quantities—"vast" in this context meaning at least 12 to 15 grams per day. Furthermore, this protein had to be "first class". It seemed, then, that we really had to consume a great deal of meat, fish, eggs, milk, and cheese—as much as possible. If we didn't, the message was, we would be feeble and (since antibodies are pure protein) we would be highly prone to infection.

There were many who doubted all this—including vegetarians, who may eat eggs and cheese (and possibly even fish, which is stretching things quite a lot) but in general consume very little animal protein. Some outstandingly able people were vegetarians, including George Bernard Shaw and Tolstoy (we will skate over Hitler). Indeed, some entire human populations were and are vegetarian to the point of vegan (no animal food at all), including many of the people of Southern India and rural Japan. Yet, demonstrably, vegans commonly live long and healthy lives. Indeed, at the same time as we were being told that people needed vast amounts of meat if they were to survive at all, we were also being warned that people who for the most part ate little or no meat were breeding too fast.

Partly because of such general doubts, and partly because of some more critical experimental studies, it became clear by the 1970s that human protein needs had been greatly exaggerated. We don't need to eat 15 grams a day. Healthy adults seem to get by perfectly well on five grams or less. Neither do we need all the protein to be "first class". Few allegedly second-class proteins are *so* deficient in essential amino acids as to be positively useless. Besides, different "second class" proteins tend to complement each other. So it is that cereal proteins tend to be low in lysine while pulse proteins (beans and so on) are high in lysine. So cereals and pulses together provide first class protein. So the cereal-plus-pulse theme runs through all the great cuisines: rice and soy in China; rice or chapatti (wheat) and dhal in India; rice or couscous with beans (often broad beans) in the Middle East; tortilla (maize) with frijoles (kidney beans) in Mexico; and even beans-on-toast in Britain. People in traditional societies may not be versed in the niceties of biochemistry but they know what to do to stay alive. If they didn't they'd be dead.

The economic, social, and overall biological consequences of this shift in attitude to protein can hardly be overstated. In the days when we were supposed to need boundless protein, it seemed essential not simply to raise livestock, but to raise as much as possible. The intensive livestock systems including the battery cage for poultry, the factory farm for pigs, and the units for raising beef on "surplus" cereal (so-called "barley beef") that were just coming on line seemed to have arrived in the nick of time. Animal lovers who objected to such intensive farming on grounds of welfare and aesthetics were held to be irresponsible. Current mythology had it that we couldn't feed the human race without intensive husbandry. Contrariwise, cereals and pulses, including soya, were presented as "stodge", to be grown primarily for animal feed.

Once the penny dropped—that actually, people don't need vast amounts of protein, and certainly don't need vast amounts of specifically animal protein—the whole picture reversed. Clearly, people could get all the protein they need from cereals and pulses alone: and while they were getting their protein, they would also get most of their energy, with a liberal dose of fibre too. Livestock that fed on cereals and pulses was *competing* with us, since we could perfectly well have eaten what they were eating. To be sure, meat and other animal products are highly desirable for a whole host of reasons. They provide essential minerals such as calcium and zinc that are not easily obtainable from plants in sufficient amounts, and vitamins such as B_{12}, and some essential fats, and are certainly significant as sources of calories—especially in latitudes and in seasons where plants are not readily available. So don't write them off. But don't build the whole diet around them either.

So the shift in nutritional theory—towards much less protein, and particularly less animal protein—could and should have transformed the face of agriculture. It should have halted the frantic emphasis on livestock. But of course it did not. For reasons discussed in chapter 4, livestock can be highly lucrative: and where lucre leads these days, all human endeavour, including agriculture, is bound to follow. Of course the food industry claims that the livestock industry continues to flourish (although not everywhere) only because it meets "consumer demand"—but this, as we will see, is far less true than industry likes to pretend, and indeed in large part is a simple lie (for money justifies untruth as well).

So, even though the shift in theory meant we no longer needed to emphasise livestock, the cash-wagon has rolled on. Currently we feed 50 percent of the world's wheat and barley to livestock; 80 percent of the maize; and well over 90 percent of the soya. By 2050, on present trends, when the human population numbers nine billion, our livestock will be consuming enough good grain and pulses to feed another four billion—roughly equivalent to the total human population in the early 1970s when the United Nations held its first World Food Conference in Rome to discuss what it saw as a global food crisis.

There is a broader lesson in this, too. If nutritional science had shown that human beings do need more and more meat, then this would have been seized upon as justification for a bigger and bigger (and more and more lucrative) livestock industry. Since modern science shows the precise opposite, the relevant science is simply ignored. Thus we see the role of science in modern politics and commerce in general. Science at its best and in its unadulterated form aspires to discover objective, factual truth. It would be safer, you might think, to base political strategies on objective factual truth where this is possible. But when the factual truth is inconvenient to the powers-that-be, they simply ignore it, or find some tame scientist who will say whatever he or she is paid to say. But I will come to that.

The last broad category of essential foods is or are loosely classed as "micronutrients": essential to be sure, but required only in minute quantities—typically in milligrams or even micrograms per day, or at most only a few grams. Micronutrients may be considered under three headings: minerals, vitamins, and a newly identified and little understood collection of what for convenience I will call "paravitamins".[1]

---

[1] *Three novel nutritional terms have become fashionable in recent years: "functional foods"; "nutraceuticals"; and "phytonutrients". Functional foods are food of any kind that seem to have some beneficial effect over and above their nutritional content. That is, they seem in some way to function as tonics or medicines, and not simply as sources of energy or raw material. The term "nutraceutical" is commonly used synonymously with "functional food". "Phytonutrient" strictly speaking should mean anything nutritious found in plants, but in practice (since most nutraceuticals are plant-based, and since bacteria and fungi are commonly classed as "plants" although they very obviously are not), the term "phytonutrient" is also used synonymously with "nutraceutical". All these terms are vague, however, and used in various contexts. I am coining the term "paravitamin" to describe the active ingredients that the various "functional foods" (alias nutraceuticals alias phytonutrients) apparently contain. I would not bore you with this except that all these novel terms are heard so often these days, not least because functional foods are now such big business. Again, we find that there is some excellent science behind the general idea: but once ideas become commercialized, as all ideas seem bound to do these days, they also become subject to spin, and confusion is bound to reign.*

Minerals are chemical elements: non-metals such as iodine which is a component of the thyroid hormone thyroxine; and metals such as sodium and potassium which are essential to maintain the integrity of cell membranes; iron which is a key component of hemoglobin; calcium which has all kinds of functions (in addition to its role in the structure of bones and teeth); and so on and so on. One way or another the body makes use of and therefore has need of about a third of all the elements in the periodic table. Some of those elements (probably most) are vital in small amounts but toxic in larger amounts. Copper is an obvious example. Sodium too is essential—but if taken in very great amounts over a long period it predisposes to high blood pressure, aka hypertension, which in turn predisposes to stroke and increases the risk of coronary heart disease. It is rare to consume too much copper but excess sodium intake has become usual in the western world because we eat so much salt—sodium chloride; not so much as a condiment, but as a major ingredient of processed foods, freely deployed both as a preservative and to enhance the flavor, including sweet foods such as ketchup and some breakfast cereals.

Our need for vitamins began to become apparent at least by the seventeenth century, when sailors and doctors realized that scurvy was caused by nutritional deficiency—and could be countered by eating citrus fruit such as limes. By the nineteenth century it was clear that the essential ingredient was vitamin C—alias ascorbic acid. Ascorbic acid, it is now known, is one of the body's many "anti-oxidants". We rely on oxygen for respiration but it is chemically lively stuff and if it escapes in various chemical forms into the body at large it can oxidize the body tissues themselves, with huge damage. Several vitamins (and other bodily ingredients, such as uric acid) are now known to function as anti-oxidants—and we would be sunk without them.

Throughout the nineteenth but especially in the twentieth centuries more and more such vitamins were identified, all very different chemically, all essential, all leading to disorder that could be fatal if present in too small amounts. Mostly they are known by letters, such as vitamin A—deficiency of which leads among other things to dryness of the eyeball ("xerophthalmia") and hence to blindness. Relative deficiency of folic acid in early pregnancy (which effectively means pre-pregnancy) apparently increases the chances of spina bifida in the developing fetus. And so on.

"Paravitamins" I am defining as the active ingredients in "functional foods" or "nutraceuticals", as described in the previous footnote. If paravitamins are lacking, this does not necessarily lead to overt disease—which is why it took so long to identify these agents. But if they are present then in various ways they seem to be health-promoting. One of the best known is or are various plant sterols which, when present, apparently lower blood cholesterol and so, in theory at least, should reduce the likelihood of coronary heart disease.

Some hard-nosed sceptics doubt the validity of paravitamins. They find it implausible that the body should require such odd and unrelated things—generally components of plants. Besides, because deficiency of paravitamins does not generally lead to obvious disease or death (in the short term), it is very hard to measure their effects; and many scientists, of a certain type, dismiss phenomena that are not easily measurable and so seem unreliable.

But as the great twentieth-century Ukrainian-American biologist Theodosius Dobzhansky commented in an essay in 1973, "Nothing in biology makes sense except in light of evolution"; and if you look at paravitamins in an evolutionary light, they make perfect sense. So it is that our ancestors—our first human ancestors, and before that our australopithecine and then our apish ancestors—included a great variety of plants in their diet. Modern hunter-gathering people commonly consume a hundred or more different species. Until the past few thousand years, all the plants that people ate were wild—and wild plants include a great many potent biochemical agents, many of which the plant produces in order to ward off insects and other parasites, and for many other purposes as well. Thus, our hunter-gathering and pre-human ancestors were exposed, every day, to a huge variety of different chemicals produced by plants for all kinds of purposes, many of which were toxins. In addition, in nature, we are exposed to a great many bacteria and fungi, ever present on food in one way and another: and, biochemically speaking, bacteria and fungi are just as accomplished as plants.

Animals cope with weird substances in many different ways. If they consume what is all too obviously poisonous, they die (although most, like rats, have clever ways of sampling and rejecting novel foods before they consume too much). Many herbivores avoid trouble by eating only small amounts of any one plant: goats and ostriches do this. Others develop specific enzymes and extensions of the gut to help them overcome specific toxins that they encounter in large amounts—and so the koala has a caecum (a blind extension of the gut) which is packed with bacteria and protozoa, which break down the many fierce toxins, resins, and fibre in the leaves of its favored eucalyptus.

But evolution, most generally, leads to adaptation. An animal may first evolve the means to cope with some plant toxin, for example by developing some detoxifying enzyme. At first, the body merely excretes the breakdown products produced by the detoxification. But as the generations pass the body finds ways of utilizing the breakdown products of the original toxin—and these may then function for example as anti-oxidants (and it is hard to have too many). So we can imagine that a body comes eventually to depend upon weird agents that are present in nature—mostly in plants, but also produced by other creatures such as bacteria and fungi—and might originally have been toxic. This, after all, is precisely how early bacteria came to terms with oxygen itself—first poisoned by it, and then becoming reliant upon it.

I suggest, then, that our need for a variety—perhaps or even probably a huge variety—of weird substances that are present in nature but are hard to identify or pin down, is entirely plausible. Indeed I have developed the concept of "pharmacological impoverishment": the condition that animals or any of us are in when our diet is deficient in these agents (see "Functional Food and Pharmacological Impoverishment", in *Future Food*, Caroline Walker Trust, London 1999). The term "nutraceuticals" is not entirely inappropriate since these recondite agents have some characteristics of food and also some characteristics of drugs, or at least of tonics. (But I prefer the term "paravitamin" for reasons described in the footnote.)

I also suggest that on modern diets, we are extremely likely to be pharmacologically impoverished. We consume nothing like the range of plants that our ancestors ate—and that we are presumably adapted to; and most of the plants we do consume are domesticated, bred over many generations primarily for yield and appearance, so that much of the biochemical variety and subtlety has been bred out of them.

What I do find objectionable, strange, and deeply pernicious is the modern approach to paravitamins. The obvious lesson is that we should eat, as our ancestors ate, a huge variety of plants—and especially wild plants; and fungi, too, and fermented foods. We need to acknowledge that it is logically impossible to identify all the paravitamins our bodies might need: only by eating many different things can we be reasonably sure of covering all bases. So the nutritionists should be saying: "Variety, variety, variety". Some of them are saying this. But many, including many who are paid the most and have the biggest laboratories to work in, are seeking instead to identify particular paravitamins, and where possible to synthesise them in the laboratory, and then add them to processed foods. They do this not because it is nutritionally sensible and good for human beings, but because it is profitable. If people were simply encouraged to grow herbs, and sometimes where feasible to gather them, there would be no profit for food processors. So instead we must be told that the only way to obtain ingredients that are essential to us is by buying particular foods produced by particular companies with particular bands of shareholders to answer to, at huge cost. This is the way of the modern world: not to do things that are merely sensible and beneficial; but to do those things—and only those things—that bring profit to big companies and at the same time increase the power of the political parties who are financed by those companies. Game, set and match. Yet, biologically speaking—which is what really matters—those huge dominating companies are redundant. This is the nonsense we have to escape from.

It is possible, of course, to write forever about the intricacies of nutrition: biological, historical, political. But the above, I suggest, includes everything that everyone really ought to know. Of course it is complicated—at least in detail. The powers-that-be revel in the complexity. The more complicated it appears the more it is inaccessible—and the more we are apparently dependent on experts; on the powers-that-be, who alone are able to handle the knowledge. Obfuscation and esotericism have been the con trick of charlatans through the ages.

Far more striking, however, and far more important, is the underlying simplicity of modern nutritional theory. For when you boil it down, what does it amount to? It can be summarized in nine words:

"Plenty of plants, not much meat, and maximum variety".

That's it. All the thousands of textbooks and diet books and healthy eating books that occupy miles and miles and miles of shelf-space in hundreds and hundreds of libraries and bookstores can be expressed in this one brief adage: "*Plenty of plants, not much meat, and maximum variety*". That's *it*.

Let us leave it there, then, and ask the next biological question. How can we best produce the food that we really need?

A lightning course in good farming

Farmers, when they are not just farming for a hobby and for fun, seek *efficiency*. But efficiency is a slippery concept. In general it means how much you get out, relative to how much you put in. Maximum efficiency implies the greatest possible output for the least possible input. But whether or not a system is considered "efficient" depends very much—entirely in fact—on which inputs, and which outputs, you are measuring. So it is that traditional farms, making no use of artificial pesticides or fertilizers, and using only the muscle power of people or animals (typically oxen or horses, but also donkeys, camels, llamas, even elephants, and so on), in general produce about 10 kcalories of food energy for every one kcal of energy expended on the cultivation. But in modern industrialized farms the equation is typically the other way around: 10 kcals are expended, largely in the form of fossil fuel, for every one that is created in the form of food energy. In terms of energy out *versus* energy in, therefore, the traditional systems are about 100 times more efficient.

But in countries that practice industrialized agriculture fossil fuel has been cheap while labor and the land required to support working animals, has become extremely expensive. Modern farms may employ a hundred or a thousand times fewer workers per hectare, as traditional farms; yet, as the fields are ploughed deep with modern tractors and the crops are plied with modern fertilizers, the yields per hectare may be 10 times as great. In cash terms, then, modern systems may be tens or hundreds of times more efficient than the traditional systems. To people who think only in terms of cash—as, by and large, the modern powers-that-be do—it is therefore self-evident that modern systems, are far more "efficient", and should replace traditional systems as quickly as may be arranged. Indeed, in official high circles in all contexts—the European Union; the World Bank; various branches of the United Nations—the industrialization of world farming is seen as a prime desideratum. Indeed, the industrialization of agriculture has become a principal yardstick of "progress".

British powers-that-be are wont to claim that Britain's agriculture has long since become "the most efficient in the world". It is, after all, industrialized to the nth degree. Indeed, modern British and American commercial farms are an exercise in heavy engineering and industrial chemistry, nowadays abetted by biotech. Only a few die-hards buck the trend: organic farmers, who cling to values other than cash and are mercifully finding now that people will pay more for extra quality (more of this in the last chapter), plus a dwindling residue of old men (mostly men) who are living on their capital (the average age of Britain's full-time farmers is now about 60).

Overall, not much more than one percent of Britain's workforce now works full time on the land—compared to an average of 60 percent in India, and in the Third World as a whole. The US is very like Britain: Thomas Jefferson, perhaps the greatest of its founding fathers, envisaged the US as "a nation of small farmers" but now there are more people in US jails, than are working full-time on the land. Yet, in cash terms, British farming almost holds its own: the one percent of farm workers produce 0.7 percent of our Gross Domestic Product (GDP), which means our farms are almost as efficient, in crude cash terms, as our factories and our hairdressing salons. This doesn't stop senior people in the treasury wanting to put a stop to British farming, just as they closed Britain's coal mines, since it would still be cheaper as things stand to buy in all our food from countries with more sunshine and even cheaper labor, even if it does mean that it all has to be flown across the globe.

But British- or American-style agriculture is "efficient" *only* in cash terms—and the cash price that we attach to inputs is entirely arbitrary, dependant entirely on historical circumstances. Oil these past few decades has been absurdly cheap. At least until the 1970s when the oil-producing countries ganged up to form OPEC, the price was absurdly low. Even in the past 30 years the "developed" world has burnt oil as if the supplies could last forever: only in the past decade or so has the word got around that no new, huge oil-fields have recently been discovered, or are likely to be, so now the oil is running out (it always was running out, but now it's official). Collateral damage has rarely been costed, and has never been costed properly. Industrial farmers have rarely paid for the rivers and offshore reefs that they have ruined, or for the fisheries that their endeavours have wrecked in passing.

Hundreds of millions of people have been thrown off the land in the cause of cash efficiency but whereas in western countries the former farm workers and their families have largely been absorbed by urban industries (where "industry" is defined generously, for many urban jobs are menial in the extreme) in the Third World they have mostly served to swell the billion-strong urban slums. The growth industries in Africa have been in mercenary soldiering for wars that are typically unfathomable (tribal feuds and foreign interests and general gangsterism and religious quibbles buoyed by perfectly justified discontent) and in prostitution. Prostitution and especially trafficking, from Africa to Western Europe (and of course from Asia and Eastern Europe) has become big business: it would be classed as a "major industry" if the traffickers paid taxes. Prostitution of course exacerbates AIDS. All in all, it's amazing how much misery can be traced to our cavalier approach to farming. But the human misery has not been costed, and if it had, the cost would not have appeared on the balance sheet of the new plantations, of oil-palm and sugar and soya and cotton and coffee and so on and so on, that have been replacing the traditional farms, or in the audits of the processors and supermarkets, who drive the whole thing along. Worldwide, the oil burnt on modern farms, converted to carbon dioxide, and the methane generated by the ruminations of the ever increasing cattle, contribute significantly to global warming. Again, this has not so far appeared on the balance sheets of modern agriculture—although if global warming were to be costed thoroughly, it would make nonsense of all balance sheets, in all industries.

This is the crux. Success in modern agriculture, like everything else in the modern world, is gauged entirely by cash, as if cash was a perfect mirror of reality. But of course it is not. Cash can be generated constructively but it can also be generated just as effectively—and often far more effectively—by means that wreck the world. By now, as human misery mounts up, and other species die out, and soil disappears and the lakes dry up and the climate grows warmer and more violent, we might have hoped that even the powers-that-be would realize that money alone is a very dangerous yardstick by which to measure success in any endeavour. By now we should be seriously wondering how and where we will be able to produce our food in 50 years time—or whether indeed we will be able to produce it at all, when the resources have dwindled even further and the climate is breaking all records. Absolutely not should we be thinking, as some influential members of the British government have lately been thinking, of running agriculture down, to save a few bucks. If there is any zeal for British farming these days it is for fuel crops, not to provide food but to sustain the automobile. We can look forward to coast-to-coast willows and rapeseed. More broadly, agriculture as a whole must be judged by criteria quite detached from present-day cash: criteria that have to do with morality and justice and with bedrock biology; with matters not of personal wealth and the mirage of "economic growth", but of survival. We need, in short, to get serious. We need to recognize that agriculture is not just a way of making money, like sports cars and hairdressing. It is what we need to do to stay alive, and to keep the world habitable; and we are in serious danger of messing it up, irrecoverably.

In short, if we take farming seriously—as we should always have done, and certainly need to do now; if we acknowledge that its central purpose is to feed people, without wrecking the rest of the world; then we need to measure its efficiency not by the flexible yardstick of cash, but according to the bedrock criteria of biology.

So: If we are to maximize *biological* efficiency, what should we be growing, and how and where, and who should be growing it? The answer to this, as we will see, is very different indeed to the answer we get if we think merely in cash. Since the modern powers-that-be *do* think exclusively in cash, and openly deride all other approaches, we also find that the agriculture we would design if we were serious about feeding people and keeping the world tolerable is different in almost every respect, structurally and in detail, from what the corporates and the world's most powerful governments are now imposing upon us. In short, our task is two-fold. First, we have to devise agriculture that really can feed us. But also, we have to by-pass or otherwise neutralize the present-day powers-that-be. The second of these two tasks is the harder, and I will discuss it in the last chapter. For now let us focus on the prime requirement—of creating farms that meet the criteria of biological efficiency.

Biological efficiency

"Maximizing biological efficiency", put more straightforwardly, means producing as much good food as possible per acre or per hectare, by means that are minimally destructive—preferably, indeed, by means that leave the soil and the waterways better than we found them. From all that has been said so far it's clear that we can reasonably focus on the macronutrients—energy and protein; and if we also strive for maximum variety we will take care of the essential fats and of the micronutrients in passing. So how do we do this?

Well, in the 1950s, when the zeal for protein was at its height and nobody thought seriously about dietary fibre (the term "fibre" hadn't even been coined) nutritionists were for the most part telling us that the central task was to maximize livestock: meat, eggs, and dairy. Now, as we have seen, the nutritional emphasis has shifted. Now it is clear that we can get all the protein we want, as well as the bulk of our energy, from plants: more particularly from the crops that are generally called "staples"—cereals, pulses, tubers, and also various oil seed crops which can be major sources of calories. Grow these in sufficient quantities, in the right places (the places where people actually are) and the problem of feeding people is all over bar the shouting. Feeding people, looked at in these simplest of terms, really is easy.

Even in the present, meat-orientated world, the statistics vindicate this generalization. By far the most important staple crops are the cereals— basically, the big, nutritious seeds of certain grasses: wheat, rice, and maize (which the Americans call corn), barley, rye, oats, sorghum, millet, and teff, the local grain of Ethiopia. The pulse crops are the beans— soya, the various kidney beans, and broad beans; peanuts (alias groundnuts); chickpeas; pigeon peas; lentils; and peas. Some other non-grass seeds also serve as grains: quinoa and amaranth from South America; "wild rice" from North America. Oil seed crops include rapeseed (known in the New World as canola), which is grown the world over; sunflowers (which prefer warm climes); olives (primarily in the Mediterranean); coconuts, and more and more palm oil (in the tropics). Peanuts, maize and soya also serve as significant oil seed crops. Nuts also are seeds and can be very important locally—notably, these days, the coconut in southern India and southeast Asia. The world's most important tubers are the potato, plus cassava, yams, taro, and sweet potatoes.

Of all these staples, by far the most important worldwide are wheat, rice, and maize. Between them, directly or indirectly (after conversion to meat by livestock) they provide humanity with a half of all our calories, and two thirds of our protein. The statistic is astonishing, and yet it is the case. But the other staples are important too. Barley is used largely for animal feed and brewing these days but is an admirable food crop in its own right with particular value where the land is salty and on high mountains (it replaced wheat in ancient Mesopotamia as the Euphrates silted up and the land was salinated and nowadays is much favored in Tibet and Nepal). Rye and oats withstand extremes of climate. Sorghum and especially millet are crops of extreme dry land—and peanuts even more so. Coconuts are wonderful by the sea, drenched in brine and sometimes seeming to grow out of pure sand, tethered by their astonishingly tough and prolific roots. Potatoes used to be written off as the great no-no. Now it is clear that they can be an adequate source even of protein, at least for adults, and for many people they are the greatest single source of vitamin C.

In short, all we really need to do to ensure that the world is at least adequately fed—that people can at least get by—is to grow staple crops in the places where they grow best. They are the priority. Staple crops in general are grown on the "field scale": the ground is ploughed and the seeds (or sometimes small tubers) are planted en masse, thousands or millions at a time. Then they are harvested en masse, typically these days by machines (notably combine harvesters). Such field-scale agriculture is called "arable", from the Latin word for "ear", as in ear of corn.

However, although staples are undoubtedly the priority, and although we would at least stay alive if we had enough of them, they do not by themselves provide a complete diet. An all-staple diet would in general leave us short of some essential fats; some minerals (such as zinc); some vitamins; many paravitamins; and a further boost to the quality of the protein would probably be no bad thing despite my earlier comments—just to be on the safe side. An all-staple diet would also be rather tedious, although many people in the history of the world have lived almost exclusively on one staple or another: many Asians on rice; the poor Irish and western Scots of the early nineteenth century on potatoes; many Scots almost entirely on oats; and so on. So we need two more classes of agriculture to run alongside or amongst the arable. These are horticulture; and pastoral.

Horticulture is the art, science, and craft of growing fruit and vegetables, herbs and spices. Horticulture may be practiced on a very large scale and sowing and harvesting (or picking) may be highly mechanized but nonetheless, in principle at least, the plants are tended individually; the word "horticulture" derives from the Latin *hortus* meaning garden, and is indeed appropriate. Horticulturalists who specialize in food crops are often called "market gardeners". Fruit specialists are commonly called "growers". Horticultural crops are valuable for their essential fats (especially the kinds that occur in leaves); for their micronutrients; their fibre content; occasionally for their energy and protein—as in avocados; and for their flavor.

Horticultural crops and staples between them can provide a diet that is excellent both nutritionally and gastronomically as many a healthy and cheerful vegan bears witness to—including many a traditional Asian, in India, China, and rural Japan. Spices, herbs, and various fermentations as in the Chinese and Japanese soy and the Japanese miso certainly help both in flavor and with paravitamins.

Pastoral farming, from the Latin *pastor* meaning care, is the art, science, and craft of raising livestock—though in truth in many modern intensive units the benighted beasts receive very little care.

From first sight, you may conclude that pastoral farming is a waste of time, effort, and space: that livestock are simply a drain on the world's resources. Many a vegetarian has argued thus. After all, human beings can live very well indeed on an all-plant diet. Indeed in many respects vegans are particularly healthy—largely or almost entirely avoiding the modern major killers of coronary heart disease, and various cancers and diabetes. Furthermore, if the aim is to produce maximum protein and calories per hectare, then plants seem a far better bet. Yields of crops vary enormously from region to region: a tonne of grain per hectare of sorghum in some dry African field may be perfectly respectable, while a highly-mechanized arable farmer in Britain's East Anglia would be disappointed with less than twelve metric tonnes of wheat per hectare. But in either case, the yield of protein and calories per hectare is generally between five and ten times greater than it would be if the same field were used for cattle. Cattle need to eat about 10 grams of plant protein for every one gram of meat or meat protein that they produce (and they drink staggering quantities of water as well).

Despite all this—that livestock don't seem to be nutritionally vital and seem tremendously inefficient—pastoral farming is very important indeed. The oft-bruited generalization—that we could most easily feed the world if everyone was vegetarian—is simply not true. To be sure, it would be easier to feed a world full of vegans than a world of hamburger and fried chicken addicts. In absolute terms, it would be eminently possible to feed the former, and it is already obvious that we cannot cope with the latter. Nonetheless: there is no system of all-plant agriculture that could not be made more efficient, in biological terms, by adding in a few livestock, provided they are of the right kind, and are kept in the right numbers, in the right ways. The trouble begins, as always, only when farmers (or the corporates and governments who make farming policy) stop thinking in terms of biological efficiency and long-term possibility, and think only of cash.

For instance: In many parts of the world, at least in some seasons, it is very difficult to raise crops at all. Arable is all but impossible when the land is too high, steep, cold, or wet or if it rains too much in the season when the grain should be ripening. Horticulture is possible almost everywhere if you invest enough in it but may languish for lack of water. But animals of one kind or another muddle through anywhere—living as camels and goats may do on the most meagre of leaves that poke between the thorns of desert trees, or as reindeer do on lichen, or as long-wooled sheep and shaggy cattle do in British hills on the coarse grasses that grow between the heather; and in times of drought or the depths of winter there may be nothing to eat at all except for beasts that fattened in better times. In all countries, too, the manure of livestock has been a prime source of soil fertility—and still should be. The dust-bowl of the American prairies in the 1930s would surely have been less dramatic (so many agriculturalists surmised) if the soil had been enriched with manure, and had not been surrendered so absolutely to grain. In many tropical countries the dung of cattle serves both as fuel and as the "daub" with which to build houses of a kind that can endure for many decades (and centuries if the walls are plastered). For good measure, for people worldwide, cattle are significant transport—as of course are camels and horses; and in many countries horses are also a source of milk (fermented in central Asia to make *kvass*) and of meat (as in France, Belgium, and Switzerland. Horse meat is very good). All in all, then, we must take livestock seriously—even though, in the modern industrial systems, geared to the maximization of cash, they are subject to such cruelties and produced so profligately.

Looked at ecologically and agronomically, livestock can usefully be divided into two main categories: the specialist herbivores, and the omnivores. The specialist herbivores are able to digest cellulose—the stuff of which plant cell walls are made. At least, they do not digest it with their own gut enzymes, but they maintain vast armies of bacteria and protozoa in their guts which digest it for them, and the animal then absorbs the organic acids that result. The animal then converts these acids into sugars. Ruminants, which in this context means cattle, sheep, goats, and deer; and pseudo-ruminants, which are the camels, llamas, guanacos, alpacas, and vicunas, harbour their helpful microbes in the fore-gut—in a vast stomach known as the rumen. The "hind-gut digesters"—horses, elephants, rabbits, guinea-pigs—keep their symbiont flora and fauna in a diversion of the hind-gut, known as a caecum. Of all these herbivores, worldwide, cattle and sheep are by far the most important, but all the rest are locally important too—guinea pigs in Peru and rabbits not least in China (and Malta).

For animals that can digest cellulose (with microbial help) it is their most important source of energy, usually by far. A cornucopia is thus opened up to them, for cellulose is the most abundant by far of all the organic macromolecules in nature: there must be trillions of tons of it out there. We, straight-gutted humans, cannot digest cellulose. But we do eat the animals that can. So via cattle, sheep, and all the other specialist herbivores, we too can partake of nature's most generous feast.

Pigs and poultry, by contrast, are omnivores. Potentially, they eat anything. Indeed they eat the same kinds of things as we do, except that they have lower aesthetic standards (and both pigs and poultry to some extent seem to be able to derive at least some energy from cellulose, and may supplement their diets with grass[2]).

---

[2] *Wild pigs have long hind guts and can derive significant amounts of energy from cellulose. Modern pigs have been bred with short hind guts and need rich food— mostly cereal—just as people do. We should reverse the trend of modern pig-breeding, and develop new strains of grass-feeders. The Berkshire is among the existing types that do well on a diet high in grass. But the Berkshire is not considered "economic" and has become a rare breed.*

Thus the agronomic roles of the various classes of livestock define themselves. Cattle, sheep, and goats (and other specialist herbivores) can feed on the kinds of vegetation that we cannot feed on—notably leaves, both grass and browse (the leaves of trees) which is especially important in the tropics, not least in systems of "agroforestry". The specialist herbivores thrive in the kinds of territories where we cannot readily raise staples or practice serious horticulture—notably in cold (or hot) hills and semi-deserts. Pigs and poultry can be fed on leftovers (pigs and chickens traditionally fed on swill) and/or on crop surpluses, or crops of inferior quality. We should produce surpluses in most years: yields of crops vary from year to year for reasons beyond our control (notably climate) so we should generally aim for more than we really need to be on the safe side, and what we generally aim for we should usually achieve. Both the herbivores and the omnivores can readily be slotted into crop rotations—cereals and other arable crops for a few years, followed by a grass-and-clover "ley" with grazing cattle and sheep for a few years, then back to arable again. Pigs and poultry are fitted in as and when they can be.

Guided by such principles—principles rooted in elementary biology—the structure of the farm defines itself, too. Of course, no two farms are exactly alike. Traditionally, some of those up in the hills of Britain were devoted almost entirely to sheep; others in the pampas of Argentina were almost all cattle; and so on. Yet, traditionally, *most* farms worldwide had a similar overall structure. All marched to the drum of their local ecology and although landscapes and climates vary enormously over the globe the fundamental principles of ecology are the same, just as the laws of physics are the same.

Traditionally, farms were generally "mixed". There would be some arable—with several different species of grain, pulse, and other staples; and, often, several or many varieties of each. There would be some horticulture: at the very least, traditional farms all had cottage gardens. Both the main classes of livestock—the specialist herbivores and the omnivores—would be fitted in as required, with the cattle and sheep feeding both on permanent pasture that could not be ploughed for cereal, in wet meadows and/or in the hills, and spending a season or so in the arable fields between crops. Two such farms I have seen at first hand come to mind. One was in China, in the mid 1990s. The entire landscape was basically devoted to rice, the flooded paddies stretching from hill to hill.

But of course there was higher ground; and on the higher ground the people grew every kind of vegetable (with yams conspicuous). Between the young rice stalks squeezed flotillas of ducks, feeding on the algae and the myriad invertebrates that bred in the paddy water. I didn't see fish, but rice farmers typically, and traditionally, raise grass carp—a virtually vegetarian species that likes tropical waters. In the village there were pigs and chickens in the road, feeding on whatever they could find. Perfect. By contrast—and yet essentially similar—I recall one of the first English farms I ever saw, in the 1950s. Dairy cattle were raised on temporary leys between the arable crops largely for butter, and also on permanent hill pasture; and—very modern for its day!—young pigs who were fed on surplus corn were given a boost in growth with the whey left over from butter-making. All very neat, and biologically impeccable.

In short: farms that are designed with sound biology in mind—with respect for the physical needs of human beings, and of the crops and livestock, and the restraints of landscape and climate—produce plenty of plants, some but not much livestock, and great variety.

And here lies a wondrous but obvious serendipity: that the output of farms that march to the drum of sound biology exactly matches the nutritional needs of human beings as defined by modern nutritional science: *Plenty of plants, not much meat, and maximum variety.*

Yet, you will very properly protest, food isn't just a matter of nutrition. Food is about flavor, texture—gastronomy; and gastronomy is at the heart of all cultures. People in Britain in World War II were well-nourished, measured objectively, but they felt deprived nonetheless and rushed to embrace a more interesting diet as soon as rationing stopped in the 1950s. Laboratory rats are impeccably nourished on patent laboratory rat-feed but that doesn't mean they are happy—and besides, people are not rats. Mere nourishment in short, mere sustainability, aren't enough.

Indeed. This brings us to the second great serendipity:

The future belongs to the gourmet

Great chefs are extremely well paid these days, and very properly. They cook some wonderful things and generally speaking the modern chefs stress the things that matter: fine, fresh ingredients, prepared as simply as possible (although as Albert Einstein said in a somewhat different context, "but no simpler"*)*. Great chefs also emphasise that the very finest cuisine, all the world over, is rooted in traditional cooking.

And what are the basic ingredients of traditional cooking, all the world over? *Plenty of plants, not much meat, and maximum variety.*

In short, we can't lose. Farms that are designed to feed people forever—deliberately tailored to conform to the bedrock principles of human, animal, and plant physiology, and to the demands of ecology— produce exactly the right foods in the right proportions as recommended by modern nutritionists; and these in turn are precisely what is required to produce the world's finest cooking. It would indeed be easier to cater for a world full of vegans than a world full of hamburger addicts. But it is easiest of all to cater for people who really care about food. The future, indeed, belongs to the gourmet.

You don't have to be rich to be a gourmet. Chefs (whether great or not) like to charge fancy prices for what is sometimes called "haute cuisine". But all the greatest cooking, as all the truly great chefs acknowledge, is rooted in peasant cooking. Peasants, almost by definition, are not rich. But they do have access to good, basic, traditional farming (and usually, of course, were farmers themselves). All serious cooks need is plenty of staples, a mass as various as possible of other plants in season—leaves, fruits, roots—and whatever meat, eggs, milk and occasional fish as may come their way, and they can live as well as any royalty. The idea that the modern commercial diet of animal fat or hydrogenated palm oil, salt, sugar, and miscellaneous additives is cheap and delicious is just another lie. We can't all be farmers, but we can all be serious cooks. I have seen women cooking beautifully in old Bombay although they lived with their families on the street: a little tower of nesting brass pots with a few burning sticks beneath. Apartments are being built in London these days without kitchens. Since the world is run by idiots, it is really not surprising that we are in such a mess. But they are cunning idiots. They know how to disempower, and stay in charge.

For instance: the cooking of France is properly acknowledged to be among the finest in the world—and the French, traditionally, both rich and poor, derived at least half their calories from bread (but what bread!). Provençal cooking is rich in beans. Traditional Italian cuisine, which at least vies with French, is based on pasta and beans (again the theme of cereal and pulse). The hugely various cuisines of China and India are rightly acknowledged as among the world's finest. In both, at least in the more tropical south, the diet consists largely of rice—tricked out merely with whatever vegetables are around, and bits of whatever animals and fish happen to be around at the time, and with fermented foods such as pickles and, in China, with soy sauce, and in Japan with miso. Turkish cooking can border on the miraculous: banquets created from wheat (cracked), olive oil, almonds, mint, honey, whatever fish happen to have been pulled from the shore and perhaps some goat, if one happens to have died that week (I exaggerate, but not much). Even in northern Europe, which seems so meat-orientated (northern Europe is richer than most, and has a lot of grass and hills, and it is sometimes hard to supply fresh vegetables year round) the traditional cuisines are still for the most part firmly rooted in plants. Bread again abounds—especially before potatoes became widespread, in the eighteenth century: Britons who ate well were said to be "stout trenchermen", where the "trencher" was the flat round loaf that accompanied every main meal and indeed could serve as a plate. Stews, traditionally, are packed with cabbage and turnips and much if not most of the calories comes from dumplings. Wasn't it all grossly fattening, you might ask? Well, there were fat people in olden times but obesity in general was not a problem. It was seen only among the rich and the conspicuously self-indulgent, and mercilessly lampooned. It was not universal as it has become. Traditional diets are high in fibre (staples, vegetables) and low in fat: dumplings are made with suet, but no one ate them every day. The quantities were great in time of plenty but although rich in flavor they were dilute in calories.

Taboos on meat—no pork, no beef, no horse meat, no kids seethed in their mother's milk—are generally economic in origin, though they are typically couched in religious terms. But in general, people who eat meat at all eat a wide variety: whatever is going. (Bush meat is not a good idea, however: the hunting of it spreads infection (both ways) and leads to extinction as the beasts are caught that are easiest to catch, irrespective of esculence or rarity. Even the beautiful and rare hyacinthine macaw is threatened by hunting, though it has less meat on it than a quail). But traditional cooking does make use of all parts of the beast—and some of the finest meals I have had even in gourmet Italy were of tripe (the stomach and other intestine, usually but not necessarily of cattle). Try buying tripe in a modern supermarket. In traditional cooking, meat is the centerpiece only on feast days—Christmas, Thanksgiving, St Bernadette's. Sunday, traditionally, was a mini-feast day, centered on "the roast". But in normal times meat serves only as garnish or stock. Once a good Italian cook has a good stock (chicken bones or fish heads are a fine start) it's all over bar the shouting.

Contrast, briefly, the diet as provided by the modern food industry. High in meat, fat, salt, sugar; massively calorific but not satisfying—not least because deficient in micronutrients; rich in artifices—additives—that add color, mask flavor, and act as preservatives to make provender look fresh that in truth may have hung about for weeks or months; each additive justified (if at all) by complaisant scientists on the grounds that, when given alone, it has so far failed to wreak any consistently measurable havoc among laboratory rats. No wonder poor kids whose mothers have not been taught how to cook and live on the nonsense churned out by the modern food industry find it hard to concentrate, and take to stealing cars. They are out of their heads: overfed yet malnourished; brains addled by a non-stop deluge of chemical junk. And the world's richest and most powerful governments stand by and watch it happen, and have lunch with the chief executives of the companies that make the junk (though careful to avoid the junk themselves) and then rush back to their respective parliaments to make excuses for them. It is disgusting. But the gaff is blown. The game is up. We have to run our affairs better than this; and since governments and the corporates they serve have proved so irresponsible, we (humanity) just have to do it ourselves, just as we have done for the past 10,000 years. This is the nettle that has to be grasped.

I will come back to this. First there is more ground to be laid. Enlightened agriculture, as is clear, is far more complicated than the modern kind—which simply grows the same crop over the biggest possible area, applying fertilizer and herbicide and pesticide according to the calendar (as opposed to the weather or the condition of the soil or actual state of the putative pests), following the manufacturers' instructions. Farming by numbers it used to be called, derisorily, before it became the norm. Enlightened agriculture is what used simply to be called good husbandry. It requires attention to detail. It requires good farmers, in short, and plenty of them. It cannot be practiced by one worker on a thousand hectares, as is now the ambition. Neither, as is often the current reality, can enlightened farming be achieved by one salaried worker and a score or so of immigrant transients of conveniently dubious legal standing (who's counting? Who cares?) to fill in the cracks. Good farmers are essential; people who understand the land, and crops, and livestock, and give a damn. Farmers in turn need back-up. Thus enlightened agriculture requires truly agrarian communities—the very thing that "modern" governments (not modern at all in reality, as we will see) are seeking to eliminate.

In short: if we really care about our own future; if we really want to ensure that our grandchildren have enough to eat, and live in tolerable societies, and have other species to share the world with—and that their children and grandchildren can in turn enjoy the privileges of this astonishing Earth—then we need to acknowledge that the future economy of the world needs to be agrarian. Behind Enlightened Agriculture lies The New Agrarianism.

Before we move on, though, a postscript is called for.

Postscript: Enlightened Agriculture and organic farming

You may well be wondering at this point how Enlightened Agriculture as described here differs from organic farming. Not much, is the short answer. But there are some important distinctions.

In general, no two farms are exactly alike and farming this past 10,000 years has surely manifested in many millions of different ways. But all of them can, at least roughly, be placed in one of three categories.

The first—traditional farms—make use of what there is. They do not make use of high technologies—mechanical power and artificial fertilizers or pesticides for the simple reason that until the nineteenth century they had not been invented, and they did not become commonplace until well into the twentieth century. So if we define "organic farming" simply as farming that does not partake of industrial chemistry, we could say that all farming was organic at least until the nineteenth century and most was organic, in this basic sense, until well into the twentieth. The world population when large-scale farming first began around 10,000 years ago is estimated at a mere 10 million. Since numbers were approaching three billion by the 1930s, when high-tech farming first became widespread, we can see how successful craft-based, traditional farming has been. It produced a three-hundred fold increase in human numbers since hunter-gathering days—and has contributed a great deal to the further increase, since the 1930s, when the population has doubled again. It simply isn't true, as some zealots for modernity seem to think, that farming was floundering until science and high tech came on the scene. In truth, agricultural science has achieved the successes it has only because it had such a firm—traditional—base to build on.

The second form of farming is the industrialized kind, which makes maximum use of mechanical power and industrial chemistry; and although industrial farming is the Johnny-come-lately, it is already commonly called "conventional farming". It is conventional, however, only insofar as it makes use of western methods, and is geared to the western, industrial economy.

The third form is modern organic farming. It is like traditional farming insofar as it makes no use of industrial chemistry (or at least, makes only minimal use)—but the modern organic farmer actively rejects the industrial methods: it's not that they are not available. Organic farms also tend to be more traditional in structure than industrial farms: more craft-based, and more labor intensive; and organic farmers apply "tender loving care" as a matter of philosophy. So in general ways organic farms have important features in common with traditional farms—much more than with industrialized farms. Yet it is wrong to imagine that modern organic farms are merely traditional, or are in any sense old-fashioned. They make tremendous use of modern science, often of the most intricate kind, to ensure for example that the soil is maintained in the best possible "heart"—finest texture, highest fertility, high organic content; and to explore means of containing pests by "biological" means, for example by encouraging natural predators into the crops. To a large extent the science of organic farming is ecology—and ecology is not an exercise in airy-fairyness, but is the most intricate of all the biological sciences. Organic farming is not an exercise in nostalgia, in short. It makes use of the most refined of all the biological sciences. The pity is that so little is spent on organic agricultural research. The lion's share of research money, and then some more, is spent on the comparative crudeness of "conventional" farming, and in particular on its latest scion, biotech. But then, conventional farms can more easily be designed to generate cash, so they have more to spend.

Enlightened agriculture as I envisage it is very like modern organic farming. Indeed if modern organic farming became the norm, then I would be happy to acknowledge that enlightened agriculture had arrived. Yet enlightened agriculture does begin from a slightly different ideological base. Lady Eve Balfour, who founded Britain's Soil Association in the 1940s emphasized above all the need to take care of the soil: "Take care of the soil and the crops will take care of themselves". Indeed. No argument. But although she and the other nineteenth- and twentieth-century philosophers and agriculturalists who lent their ideas and scholarship to the early organic movement included many with advanced social conscience, the organic movement was not founded specifically to provide good food for everyone forever. It could do this, probably, if well applied. But that was not its specific agenda. It is, however, the prime agenda, from the outset, of Enlightened Agriculture.

You may feel this is just a quibble, but it has practical consequences. Notably, the rules of modern organic farming banish a great many technologies absolutely, and as a matter of principle, including artificial fertilizers, pesticides, and of course genetic engineering. I would not myself ban any technology *a priori*, on first principles. Sometimes artificial fertilizers made by industrial processes can give a crop just the boost it needs to make full use of transient sunshine or rain. Sometimes particularly recalcitrant pests can be controlled most efficiently, and with least collateral damage to other wild creatures, with pesticides that are now banned. Later I will discuss some applications of genetic engineering that seem to be benign—that could truly make life easier for traditional communities who wish to retain their own ways of life. The trouble, I will argue, does not lie with the technology itself, but with the economic framework in which it is now obliged to operate, which ensures that these highest of technologists are deployed primarily or even exclusively by corporates, who in turn work hand-in-glove with powerful governments, so that in practice they become agents of social and political control— obliterating the ways of life that they could be abetting.

In short, Enlightened Agriculture would allow itself to be more catholic in its choice of technology, than organic farming is. Even so, I recognize that the high technologies that the organic farmers eschew have in practice become agents of top-down governmental control; and that in this crude political world of ours the only practical way to prevent a top-down takeover is to ban these technologies altogether. Thus, organic farming has taken on the mantle of Enlightened Agriculture by default. As things stand, in this crude and aggressive world, the "organic" label serves at least in part to demonstrate that the food in question has not been produced by means that are beyond the pale. If and when general awareness is raised, and only then, will it become possible to relax the rules without re-opening the floodgates to the present horrors of wall-to-wall monoculture and factory livestock.

But finally, and very importantly, we can already see that the "organic" label, defined in its present forms, is capable of serious abuse. In Britain we can buy "organic" apples from New Zealand, which is as far from Britain as it is possible to get without leaving the Earth. The main street of a small town near me is often blocked by a vast truck, as big as a warship, bearing organic produce from several counties. Organic crops more and more are grown as monocultures, on the vast scale. In short, organic farming is becoming industrialized just as the "conventional" kind already has been—with the same top-down control. It is important, therefore, *not* to define Enlightened Agriculture purely in terms of one particular technical approach. Agriculture becomes truly enlightened only when it keeps all the balls in the air—biological, social, moral—with the general aim of creating a world that is good for everyone forever, and for other creatures too. Unfortunately, therefore, as things are, enlightened agriculture cannot simply sneak in behind the skirts of organic farming. It overlaps the organic movement very considerably, and some organic farmers are among the most enlightened of all and are indeed exemplars for the whole world. But still, Enlightened Agriculture must have its own identity.

So where have we gone wrong?

In principle, it really should not be difficult to supply everyone who is ever likely to be born with great food, forever; and to do so without wiping out other creatures, and generally wrecking the fabric of the Earth.

In reality, present day agriculture is not directly geared to human wellbeing, and takes virtually no account of biological reality. Instead, it is designed to make money, in the apparent belief that the maximization of disposable wealth is both necessary and sufficient. How this works, why it is so destructive, and how this perverse state of affairs came about, is discussed in the next two chapters.

(After that—in the last two chapters—I will revert to a positive vein and discuss what we can do to put things right.)

Chapter 4

THE ROT SETS IN: FARMING FOR MONEY

We are failing, miserably, to feed ourselves properly. Along the way, we cause huge collateral misery while wrecking the fabric of the world itself. If we go on as we are then life will be impossible for own children and grandchildren. Why are we behaving so perversely?

The powers-that-be are in charge (by definition) and I have often asked in rhetorical vein, are they stupid, or are they wicked? Actually the question is not quite rhetorical, because some people in high places are remarkably ill-informed and demonstrably stupid, and are seriously wicked when judged by what seem to me to be reasonable standards. Indeed, the logistics of power to a significant extent favors wickedness in the highest places because the people who are most likely to acquire the most power are the ones who are most focused on power; and the desire for personal power seems largely incompatible with the primary virtues of respect for others, and of personal humility.

But many rich and powerful people are neither stupid nor wicked. Rich and powerful people include some of the most intelligent of all and some of the best: people who really do want the world to be a better place, and seek to benefit humanity as a whole. Some of them are renowned historical figures—the kind generically known as philanthropists. Such people still exist and I know some of them personally: social entrepreneurs who set up businesses (or maintain businesses that bring little profit) specifically to benefit their own societies. I have seen this in particular in India among people from both a mercantile and a Hindu tradition. In the West the Quakers, steeped in Christian morality, have been serious commercial players—and of course you don't have to be either a Hindu or a Christian to be humanitarian. So the idea that wealth and power necessarily reflect stupidity and wickedness is just plain wrong.

So the fault does not lie primarily with wickedness, or with stupidity. It lies with error. We have contrived by degrees to create a world economic system that is bad for humanity in general and disastrous for agriculture in particular—the thing we absolutely have to get right. How come?

What's gone wrong?

To people brought up in the Cold War—which must include most people reading this since it officially ended only in 1989—the world's economy is clearly divided into Communist and capitalist. The West is generally conceived to be capitalist, and those who feel that the present western economy is not what the world needs, are still liable to be called "Commies", and banned from serious further discourse.

In truth, although the Communist party still rules in China, and China is rapidly becoming the world's biggest economy, the kind of economics that Communism traditionally embraced is now rare. In this, as first envisaged by Karl Marx, the people at large own "the means of production". People-at-large in reality means the state; so the economy is centrally controlled. Only a few modern countries adhere to this model, such as Cuba and North Korea. The Communist party is still big in Russia and is the ruling party in China but both preside over economies that are unmistakably capitalist.

Many conclude from this that the ideological war is over. Indeed, all the world's economies—all except those eccentrics that have kept themselves to themselves or have been shunned—subscribe to the global free market, presided over by the World Trade Organization, based in Geneva. Capitalism has won, and that is the end of it. Some observers, including many in the highest places, heave a huge sigh of relief. All we have to do now, they argue, is to get on and make global capitalism work. The world is now united—at last! The free market delivers the goods that we all need, with wondrous efficiency. For good measure, it has a firm moral base. After all, traders can succeed only by supplying what consumers will buy, so consumers must be in the driving seat. Since everyone is a consumer, the system is innately democratic. Who could ask for more? Anyone who isn't an enthusiastic capitalist these days must be an idiot, or (as George W Bush put the matter), "evil". Doubters must be the enemies of democracy, and world unity, and therefore the enemies of humankind itself.

What this simple-minded but alarmingly common view of the world significantly fails to register is that there is a huge, deep, ideological and practical division within capitalism itself. On the one hand we have the global, allegedly free market, which now prevails. On the other hand, we have capitalism as envisaged, and espoused, by the founders of the modern United States—Ben Franklin, George Washington, Thomas Jefferson, James Madison and the rest—at the end of the eighteenth century and the beginning of the nineteenth. The difference between the two capitalist models is almost as profound as the difference between capitalism as a whole, and the centralized economies of the Marxists. People like me who feel that the global free market in its present form is a disaster are not necessarily "Commies", or religious "fanatics", or hippies or weirdos. On the contrary, I see myself as a good Jeffersonian. The United States was the greatest social experiment ever undertaken, breathtaking in brilliance and moral sure-footedness, and if only the US had continued as it began, the world would now be a very different and a far better place. I hate the present system—we have to stand up to it—but in hating it, I claim to be a better capitalist than its modern practitioners.

In truth, there seems to be no universally agreed definition of "capitalism"—all the world's biggest concepts prove remarkably elusive when you look at them closely. But I argue, as many do, that capitalism is really about free trade, markets, and personal ownership, and as such it seems to be as old as humankind. Neanderthals clearly traded in stone tools, often over long distances; and they may have traded in a great deal else besides, of which there is now no trace. Trading might almost be seen as the defining characteristic of our species. It is indeed "natural". Being natural does not make it morally right. But it does make it easy to live with.

In the beginning

It is impossible to say when modern capitalism really began—there were money-changers in the Bible, and Europe's Renaissance was financed by Italian bankers. But a key player, beyond doubt, was Adam Smith: twin pillar, along with David Hume, of the Scottish Enlightenment. In a series of essays, but particularly in *The Wealth of Nations* of 1776, he explained why the free market could indeed meet the needs and desires of humanity. In a well-tempered market there should be many different traders, all competing for customers. The customers would be free to choose between them. Traders who tried to palm off inferior goods, or cheated, would soon be found out. Then the customers would go elsewhere, and the cheats would go out of business. Thus, said Smith, an "invisible hand" would ensure that honesty and justice prevailed. In other words, each individual player in the market, whether trader or customer, merely had to do his or her own thing and, as if by magic, an efficient, honest society would result. By contrast, said Smith, if people set out consciously to design an agreeable society by imposing some particular vision upon it (as, for example, various religious leaders were wont to do) then the result could be anything but agreeable.

Adam Smith's model is alluring and in principle it surely works. But he hedged his thesis with caveats. First, he was a moral philosopher before he was an economist, and he did not envisage that the free market was society's only driver. He imagined that the individuals who took part in the market were, at least to some extent, moral beings: that they were possessed of what in his *Theory of Moral Sentiments* he calls "natural sympathy". For example, a trader was perfectly entitled to shout more loudly than his rivals. But he could not murder them. The market overall was subject to the law of the land, which in turn was intended to reflect the general moral beliefs of the people. Smith recognized too that the market, as he envisaged it, was an ideal—it could never quite be realized in practice. The invisible hand could be relied upon to dispense justice only if there was an infinite number of traders, competing on level terms; and an infinite number of consumers who each had perfect access to all the traders, and perfect knowledge of what was going on. If any trader had a monopoly, or groups of traders—or consumers—ganged together to form cartels to put pressure on the rest, or if information was concealed or misrepresented, then the invisible hand could not work its magic. The market would simply be dominated by the strongest players for their own particular benefit.

Jefferson and his fellow founders set out to create a country that was, some might say, a utopia: except that "utopia" means "no-where" and is by its nature imaginary, while the new United States was wonderfully real. The founders listed what they thought was desirable. Democracy was the sine qua non: as Abraham Lincoln put the matter a few decades later, government "by the people, of the people, and for the people". The founders wanted individual people to be free because freedom is an essential prerequisite of personal fulfillment. They wanted social justice, for as they spelled out in the Declaration of Independence of 1776, "... all men are created equal in the sight of God". They wanted efficiency, because the antithesis of efficiency is waste, and waste is a sin. They knew full well that these desiderata are to some extent in conflict. Absolute personal freedom implies action without restraint which is socially catastrophic. Democracy and economic efficiency don't necessarily make easy bedfellows. Nonetheless they felt that the market economy as conceived by Adam Smith would serve their needs. So they founded the world's first, consciously-designed democratic republic, and it was (by most definitions) capitalist.

But the founders' vision has been horribly betrayed. Jefferson must be spinning in his grave. Enthusiasts for the global free market are fond of citing Adam Smith, the austere, Enlightenment Scotsman. But they are almost as far removed from Smith as Marx was.

The end of the invisible hand

Since the founders' day there have been three key changes.

First, the market as a whole has been taken over, and is now largely controlled, by corporates. Even worse—the *coup de grâce*—the world's most powerful governments now depend on those corporates. In the US, no presidential candidate can get a sniff of power without corporate wealth behind him. Britain's New Labour party has been anxious since before it first came to power in 1997 to assure the city—the corporates— that it is on their side; that it would do nothing to inhibit them. To a significant extent, the world's most powerful governments are extensions of the corporate boardroom.

"Ordinary" companies, by and large, are each dedicated to some particular *métier*. Some make furniture, some make pickles, and so on. They need to make a profit—take in more money than they spend. But the profit is a means to an end. The end is to stay in business, so they can make more armchairs or chutney. Corporates are very big companies, or conglomerates of companies, that are not necessarily dedicated to any particular activity or product. In principle they might make furniture *and* pickles, and a great deal else besides, shifting investment from one to another according to what is most profitable. Corporates are, in fact, engines for generating cash. That does not make them innately bad. But it certainly makes them powerful—in cash, of course, but also politically, because governments need cash.

Jefferson and his fellow founders knew all about corporates. The Boston Tea Party of 1773, a key event in the build-up to independence, was a protest against the British-based and all-powerful East India Company, which in that particular instance was using its economic and political muscle to bully small American tea traders out of business. Jefferson also knew that in a free market, corporates are bound to arise. There is no innate mechanism within the market to prevent companies growing bigger, or from forming conglomerates. Furthermore, the bigger they got the more they could promote their own growth since they could use their enormous buying power to drive hard bargains and to undercut rivals. As the adage has it, money goes to money. So Jefferson and Madison in particular framed laws to restrict the power of corporates. Corporates could not, the law said, extend their influence beyond their own state. They had to renew their licences every year. They were kept on a very right rein, by the democratically elected government.

But at the end of the nineteenth century, as every American law student learns, the law that constrained the corporates was overturned. In fact, as Thom Hartmann admirably explains in *Unequal Protection: The Rise of Corporate Dominance and the Theft of Human Rights* (2004), the law was not overturned. The court case in which the overturning allegedly took place was simply misreported. But it was widely perceived that the law had been overturned and what was perceived to have happened, but didn't, in one courtroom about 130 years ago, has changed the course of the world. The liberation of the corporates was, perhaps, the most important single event of the nineteenth century, or indeed of the modern world. All the world's most powerful governments are now beholden to corporates and so, of course, is the global market. But a market with only a few giant players is very different indeed from the market that Adam Smith envisaged, with a virtual infinity of small players, none of whom had any particular influence on the whole. The modern global economy is not the dynamic, restless interplay of infinite components that Smith envisaged. It is a ponderous clash of titans. In such a system, the invisible hand that is supposed to create social justice, does not come into play.

The second huge shift has been the rise of monetization. The idea has grown—it has always been around—that money can buy anything. As Bob Geldof recently put the matter, if the people of Niger had money, they could buy food. They were starving only because they lacked money. If this is true, then it seems to follow that the greater the pile of money, the better. The creation and the generation of wealth—almost by whatever means—is thus justified on moral grounds. After all, the more money you have, the more good you can do; and if you have no money, you cannot do any good at all. Anthony Trollope summarized this notion in *The Way We Live Now* (1875). His fictional banker, Melmotte, tells his gullible, potential investors: *"These are great times and I am proud to be an Englishman in these times! What is the engine of this world? Profit. Gentlemen it is your duty to make yourself rich!"*
Trollope intended this as parody, and Melmotte is an obvious villain. But his stated philosophy seems now to be taken for granted.

The moral excuse for extreme personal wealth is provided by the notion of "trickle down": only a few may be rich today, but the wealth of the few must spread to the many. Up to a point this is true. Rich people can employ poor people, and entrepreneurs who are truly socially inspired can create fine industries that give rise to and support entire communities. There are countless examples.

But wealth doesn't necessarily "trickle down". For century after century, the rich may stay rich and the poor stay poor—the rich using their power to ensure that they maintain their ascendancy. As economists at the merchant bankers Goldman Sachs recently put the matter, "The most important contributor to higher profit margins over the past five years has been a decline in labor's share of national income." The rich grow richer and the poor grow poorer not only in all those Third World countries which the rich countries like to believe are "corrupt", but in the US itself, the world leader of the new economic order. Thom Hartmann records in *What Would Jefferson Do?* (New York: Three Rivers Press, 2004) that since 1981, when Ronald Reagan became president, the USA has grown richer and richer while the real income of the middle class has declined by 10 percent and the minimum wage of the poorest people has fallen by 17 percent. Eighty percent of American homeowners of low and moderate income now spend more than half their income on housing, and half of the ever-spiralling tally of bankruptcies are brought about by medical bills (and how many of these bills result from the diet? Diabetes for instance can be very expensive, leading as it does to horrible circulatory problems and blindness). In Britain, my children's generation is similarly obsessed, perforce, with putting a roof over their heads (and with paying off their student loans). I recall a world-weary comment from George Orwell from the 1930s, that "The average Englishman owns nothing—except perhaps a house". Only now can we see the unintended irony of his comment.

Such, though, is the faith in "trickle down" that Britain's Tony Blair and Gordon Brown apparently believe that salvation for African countries lies with foreign companies (notably corporates) setting up shop within their boundaries, and making money. Somehow, the wealth created is supposed to bring general benefit. In a recent lecture at Oxford a member of Britain's House of Lords assured us that the spanking new Manhattan skyline of Nairobi, its new business center, is the harbinger of Africa's great future. It reminded me rather of Shelley's Ozymandias, a monument to anachronistic vainglory. I have heard African leaders publicly praying that no one discovers oil in their country. Oil brings untold riches. But most of the people in most oil-rich countries have not noticeably benefited at all. Indeed, they often have the most vicious regimes of all.

But such is the emphasis on wealth per se, that governments nowadays measure their own success in it: specifically in terms of GDP—gross domestic product. Increase in GDP is called "economic growth". Nations that are "growing" fastest are deemed to be the most successful, and those with little or no growth are perceived as lame ducks or even as "failed states". Failed states are ripe for intervention, or so the rich like to think. It would be doing them a favor to take them over.

Yet John Maynard Keynes, no less, warned that GDP has very little to do with wellbeing. Wellbeing is not what GDP is intended to measure. For one thing, any kind of activity that makes money is deemed to contribute to GDP. The destruction of New Orleans by hurricane Katrina in August 2005 could well contribute in net to America's GDP because the construction industry is having a field day putting it together again (or would be, if anyone was prepared to pay). War is extremely lucrative, which is one reason why there is so much of it. Crime in the US in particular is a huge industry: all those prisons, prison officers, policemen, lawyers, clerks, cleaners, drivers, and manufacturers of safes, burglar alarms, guns, truncheons, cop cars and goodness knows what contribute wonderfully to GDP. It is a tremendous wheeze, economically speaking, to put so many people in prison (more, in the US, than are working full-time on the land). Mayhem, war, and crime hardly contribute to human wellbeing, so most sane people would surely agree. But they can be awfully good for the GDP and GDP—"growth"—dominates all political agenda.

Finally, the world is all too obviously finite. If individuals have unlimited wealth then in the end they compromise the rest of us—unless they invest all their wealth in causes that are genuinely and obviously beneficial, which is remarkably hard to do. You have to be rich to own a mansion with its own private stretch of beach, and it must be fun. But for people at large who enjoy walking along beaches, and feel they have a perfect right to do so, it is immensely irritating. I do not subscribe to Pierre-Joseph Proudhon's assertion that "property is theft". But ownership should not imply absolute right, and should not be without limits. That principle has been enshrined in most societies that we know about, since the beginning of time. But we seem to be in danger of forgetting it. Nowadays, it begins to seem, the richer people are, the more admirable. One of the Russian oligarchs who effectively stole the wealth of his entire country now owns England's leading soccer team, and is one of our modern heroes.

The third shift in emphasis is perhaps the most pernicious. In the modern global economy, morality itself is defined by the market. Adam Smith, and the United States founders, envisaged that the market would be constrained by the moral rules of society. But in the modern market, in this age of moral relativism, what is "good" is taken by definition to be whatever people are prepared to pay for. Grisly though this argument seems, I have heard it earnestly defended. The market is held to be ultimately democratic, since (allegedly) traders can succeed only by supplying what people will pay for; and what people will pay for is what they themselves feel that they need and want; and supplying people with what they need and want is what democracy means (or so the argument goes). But in reality the poorest people with the greatest needs cannot afford to buy anything at all, and if the market is all there is, then they are sidelined. Neither could the market, left to itself, produce the kind of society that most sane people would find agreeable. Many people are prepared to pay a fortune for child pornography and for child prostitutes, but no one that I know would defend either on market grounds. The abuse of children is one of the western world's few remaining taboos. But I have already heard human cloning defended on market grounds—some people want copies of themselves, and some doctors are prepared to provide the "service", so what's wrong? Similarly, the trade in guns in North America (or "personal side-arms" as the euphemism has it) is justified partly on a misreading of the constitution (individual states *in extremis* are allowed to raise a militia) but mostly on market grounds. People want guns, and will pay for them, so what's the problem?

Worst of all, the particular moral message that emerges from the modern market is that material success is all, and that material success depends on ruthless competitiveness. There are, as an entire genre of Hollywood movies never tires of telling us, "winners and losers", and winning is all that matters. The inbuilt moral restraint that Adam Smith envisaged—the "natural sympathy"—has been overridden. Business people who are truly in tune with the modern ethic do not play to the rules but instead—like professional footballers—they play to the referee's whistle. But the richest of them are so powerful, that they can use their wealth to by-pass the referee. They calculate whether it is more economical to obey the law, or to break it and pay the fine—or simply appeal against whatever judgment is brought against them and continue trading as usual until the original charge becomes irrelevant. If they are really rich and powerful, they can buy the law itself. In *The War Against Nature* Robert F. Kennedy, Jr. a distinguished environmental lawyer, son of the senator and nephew of the president, tells how he sued an American meat company whose piggeries were polluting several if not most of the rivers of North

Carolina. After a decade's endeavour, he brought the company to court. He won. Then he lost—for the company appealed, and by the time the appeal was settled, two years later, the law had mysteriously been changed to enable them to dump their ordure with impunity. To many, including many in the highest places, this kind of behavior is perfectly acceptable. Winning is the name of the game and whatever wins is good by definition. Might is right. The winners write the history. Everyday mottos capture the point.

For my part, I claim to be a good capitalist, and am sure that an economy that can truly serve the needs of humanity must be fundamentally capitalist. But we need a new model of it; far closer in spirit and to some extent in detail to the vision of Thomas Jefferson, than the world of the modern corporates. Indeed, the present world economy, this ruthless yet choreographed global battle, is a corruption, a weird extrapolation, of what capitalism is supposed to be. It is at best a diversion. We need new economic models, albeit fundamentally capitalist models; and as briefly discussed later, some people are on the case.

Nevertheless, at least for the purposes of argument, I do not want to claim that corporates are all bad, or that the globalized economy is all bad. The present system does produce some things efficiently—big airplanes and big ships; possibly automobiles and computers—and perhaps this is good for humanity. Let others argue this, elsewhere.

I do claim, however, that the present economy, the globalized clash of corporates and of corporate-dependent governments, is disastrous for agriculture.

"Agriculture is just a business like any other"

I first became aware of the turn that the economy was taking in the 1970s when I worked for the British farming magazine *Farmer's Weekly*. It was then that I first heard the chill phrase, "Agriculture is just a business like any other".

Of course agriculture is a business. I dare say—stretching a point—that Cain and Abel thought of themselves as businessmen. They produced food, I imagine on their own initiative, and expected to sell it.

That's business. Stalin tried to create agriculture run by "workers" who merely toiled for the state, with no scope for individual initiative and no special reward for special effort, and largely failed. He would have failed even more disastrously if he had not, perhaps cannily, given the peasants at least some scope to do their own thing. And peasants, at least when they are allowed to be, are business people who expect to feed themselves but also to sell their surpluses for a profit.

Of course, too, business is to some extent competitive. There isn't room in the world for all the people who would like to make things or to trade, and although we can share out the opportunities by imposing quotas (which remain a very useful device) it is nonetheless good for everybody if there is at least some way of ensuring that people who are truly incompetent, lose out to people who do the job better. Otherwise, at best, the incompetent ones simply waste precious resources.

Yet no society throughout history that I can think of, has regarded its agriculture simply as a business. Always people at large and the powers-that-be have felt that farming is in various ways special. Most obviously, its output is vital to us all, and cannot be interrupted even for a few weeks. No other industry can claim as much, so convincingly. Yet the output of this vital industry is bound to vary from year to year, unpredictably, for reasons that are quite beyond human control—the greatest single reason being the vagaries of weather. Total wipe-out of crops or livestock is of course a disaster for everyone. But glut is bad news too, at least for the farmers, for it simply sends the price right down—so that very large yields, paradoxically, can force farmers out of business. Farmers are human beings too, and it seems unjust that they should go bust when their crops fail through no fault of their own; and perhaps even more unjust if they go bust just because, in a particular year, they produce more than society needs. In the long term, too, at least in traditional societies where farming had to be labor intensive because there were no big machines or industrial chemistry, it was in nobody's interests for farmers to be put out of work, because society at large all too obviously needed them.

So although farmers have commonly worked within a general atmosphere of free trade, most civilised and organized societies have manipulated the market to ensure that the farmers, who produced the food that all of us rely upon, stayed in business. Thus societies at least since classical times have evolved a whole series of sticks and carrots, checks and balances—quotas, tariffs, grants, support buying, incentive payments and so on—to ensure that farmers, although in principle free to do their own thing, in practice do the things that society wants them to do, and do not go out of business through no fault of their own. This in principle seems both just and sensible.

At the same time, competition has always been a fact of business and a fact of life—but not the only fact. Farmers have often formed all kinds of cooperatives, for example in creating various marketing boards. If cooperatives become cartels then this is bad for the consumers, and for the producers who are not in the cartel. But cooperatives in general can be of enormous help to everyone. Individual farmers in particular regions have traditionally competed with each other, not least in the agricultural shows, when each farmer tried to demonstrate that his cow or pig or sack of barley was better than his or her neighbor's. But equally, within farming communities that are truly worthy of the name, such competition was little more than friendly rivalry, each competitor spurring the others to greater effort. When the chips were down, as indeed in traditional societies they often were, neighbors helped each other. More broadly, in my days at *Farmer's Weekly*, I was always impressed by the freedom and generosity with which individual farmers discussed their methods and their successes with the world at large. In general, if they found a good way to do things, then they wanted others to share in it. Nowadays I am proud to be at least loosely attached to the Food Animal Initiative, based in Oxford, whose raison d'être is to pioneer new methods of husbandry and to share them with farmers at large.

Nowadays, though, all-out competition is perceived as the prime and necessary virtue. The present British government, taking its lead from Margaret Thatcher's philosophy in the 1980s, is basing its entire policies in all spheres, including medicine and education, on the perceived need to compete. The very obvious fact that cooperation has its virtues too—most obviously it saves a great deal of fighting, which takes a huge amount of energy for no constructive purpose—has been written out of the act. Competition has become the order of the day, not because it works particularly well, but as a matter of dogma. The dogma dominates even though common sense and simple hawks-and-doves game theory show that societies that are essentially cooperative can most easily produce an outcome that is best for all; and also shows, contrariwise, that societies that indulge in all-out competition can never produce outcomes that are good for all but must always favor the hawks, and (though to a lesser extent) those who kow-tow to the hawks.

Our food supply is particularly badly hit by the new economics. The things that farmers need to do in order to stay financially afloat are absolutely at odds, diametrically opposed, to what needs to be done if we seriously care about the 6.4 billion people who are with us now, and the nine billion who will be here by 2050, and the other five to eight million species with whom we share this Earth; and if indeed we care about our own children. Not capitalism per se, but the modern form of it, is the greatest mistake that can be conceived. This is easily demonstrated.

How to be a successful capitalist— and why farming is different

To begin with, the aim of the universal money-game isn't simply to make money, but to generate profit: profit being the difference between the price received for the finished goods, and the cost of producing them in the first place. Profit per se is not of course bad. In essence profit is a way of keeping score. It is the obsessive desire to maximise profit without moral restraint, and to do so with maximum competitiveness so that those who are not so single-minded go to the wall, that is so destructive.

A very good business friend of mine (who had his own business, and taught in a business school) explained to me years ago that to make a profit—in any business—you have to do three things:

First: maximise turnover. The more you have to sell, the more money you can make. "Pile 'em high and sell 'em cheap" was Michael Marks's adage, of Marks & Spencer. Although you shouldn't sell 'em too cheap, because of the second requirement. To whit:

Second: add value. Adding value means by and large that you begin with raw material, or something very like it, and then you turn it into a "good" (as the economists have it) that somebody actually wants to buy.

Third: minimise costs. Pay as little as possible for the goods in the first place.

These three requirements are the bedrock. Kept within reasonable and moral bounds, they are fine. They are no more than common sense. But if you set out expressly to maximise profit, with no other end in view; if you are determined to be as ruthless as is necessary to achieve this maximization; and if you operate within and contribute towards an atmosphere of to-the-death, all-against-all competition; then the world is in trouble. Specifically, this simplistic approach produces a system of farming, and an overall food supply chain, that seems expressly designed to ensure that humanity as a whole cannot be well fed, and in which vast numbers of people are bound to be treated unjustly, and other creatures are bound to go extinct, and the future in general is left entirely to hazard. Let us take the three prime requirements one by one: maximum turnover, maximum value-adding, and minimum costs and see how this pans out.

The trouble with growing too much

In the 1970s it looked for a time as if it was going to be well-nigh impossible to feed the world population of the time, which then stood at something over four billion. In particular, yields of cereal in India were simply too low. It should have been possible to increase those yields simply by adding artificial nitrogen fertiliser ("N"). But India in those days grew old-fashioned varieties of wheat which had very long stems—that is, they were tall, like the shoulder-height cereals of Breughel's paintings. When these varieties were given extra N, they simply grew even taller, and fell over, or "lodged" as farmers say. But from the late 1960s, using breeding techniques that fell short of bona fide genetic engineering but still were very fancy indeed—involving direct transfer of chromosomes from plant to plant—breeders were able to produce a new range of "semi-dwarf" varieties of wheat, with short stems. When these were plied with extra N, they didn't grow tall. They produced more grain. The semi-dwarf varieties caught on and now most wheat worldwide is dwarfed. Modern wheat often is hardly above knee-height. But the yields of grain, when the crop is heavily fertilized, are fabulous: three times what was common in the 1970s. Comparable semi-dwarfing genes were then introduced into rice, which vies with wheat as the world's most important cultivated crop of all. The development of semi-dwarf varieties, and the heavy fertilization and irrigation (sometimes) that went with it, was "the Green Revolution". Its perpetrators received Nobel Prizes.

There was, and is, a lot wrong with the Green Revolution. For one thing it put a lot of farmers out of work, with hugely adverse consequences. It required continuing capital input, mainly for nitrogen fertiliser, so on the whole it achieved less than is desirable for the poorest farmers. But it did produce a lot more grain and at first sight, it looks like, and has often been presented as, an unequivocal victory for high tech. Before the Green Revolution, with old-fashioned varieties, people went hungry. In the immediate aftermath, for a time, India, China, and Mexico emerged as net grain exporters. Can't be bad, you might think.

But yield isn't everything. Context is all. It is good to grow a big crop, to be sure. But there is no point in this if the crop cannot be sold. If everyone achieves high yields there is glut. The market is flooded. The price falls. We have seen the truth and disaster of this in the world's coffee market these past few decades. People worldwide have been drinking coffee at least since the seventeenth century but recently it has become even more wildly popular. Traders operating on the global scale have wanted more and more. Spurred by the modern vogue for global competition they have encouraged farmers to grow more and more of it—not simply the traditional coffee growers in Brazil, Costa Rica, Mexico, Kenya and so on but in places that a few years ago had hardly heard of coffee, such as Vietnam and East Timor. Fair enough, you might think. Except that too much has been produced. Between the 1980s and the 1990s the price of coffee on the global market fell seven-fold. I have seen entire plantations of coffee on Brazilian hillsides abandoned, dying in the sun, because the farmers knew that it would cost more to harvest the crop (let alone to grow it in the first place) than they would get for it on the world market.

Zealots for global free trade driven by maximum competition seem to see nothing wrong with this. If a Brazilian farmer goes bust because a Vietnamese farmer can grow the crop more cheaply—well: that's life. "The consumer", we are told, benefits from the new cheap price—although of course that is not true: coffee on the high street becomes more and more expensive. It's the traders who benefit; the corporates who organise the global market and the governments who are supported by corporate wealth.

Advocates of the modern, allegedly free global market insist that it is rooted in the ideas of Adam Smith. Yet in practice the modern global free market has perverted the vision of Adam Smith just as profoundly as the predominant economy of the modern United States as a whole has perverted the vision of Thomas Jefferson. In practice, the allegedly free global market—in all goods, but of course including those of farming—is orchestrated by the biggest and most powerful players in a kind of global dog-fight. Nobody does well out of a dog-fight—certainly not the dogs—except the organizers. Market zealots argue that the Brazilians' bad luck in the coffee market is the Vietnamese bonanza: swings and roundabouts; it all works out in the end. But this is just another lie. Some Vietnamese business people undoubtedly do well from their new industry but the farmers as a whole certainly do not. They can sell their coffee on the world market only insofar as they can undercut the Brazilians. So they must be prepared to work even harder, for even less money. Meanwhile the organizers of the dog-fight swan around the world buying from the most desperate. Soon the Chinese seem likely to produce more coffee more cheaply than anyone else (the Chinese have every kind of climate and in principle can grow anything, and of course can invest heavily in start-up) and then—goodbye Vietnam. China has already swamped the world's textile market: Sri Lankan politicians have commented that the damage thus done to the Sri Lankan people far exceeds the destruction wrought by the 2005 tsunami (although Europe, which has economic and military muscle and so has the power to bend the global rules, has temporarily been able to keep its own textile industry afloat).

Under the present global free market, with every farmer exhorted to produce the maximum, there can never be stability again. As soon as some new agricultural enterprise is up and running, someone else, in some quite different place, with some temporary advantage—even cheaper labor is the usual one—will undercut them. Yet this system, in which only the most desperate can sell anything at all (unless they buck the system and are subsidized with taxpayers' money, as in the United States and Europe), is supposed to provide the route to universal wealth. It is obvious nonsense: either a horrible mistake, or the most hideous exercise in cynicism. I digress slightly, but the point is made. Maximization of yield is not necessarily a good thing. In the short term it leads to glut, and glut is very bad for producers—although it may well be turned to advantage by those who are organizing the global fight.

Glut, however, is merely the short-term disadvantage. In the longer term, the urge to maximise yield can produce far more lasting damage. Thus, and in particular, African farmers traditionally have not sought to maximise yield because they know that this is not usually the priority. The African climate in general is immensely fickle (and with global warming will be more so). There are very good growing years, and very bad ones. What matters is not to produce vast crops in good years (leading to glut) but to guarantee sufficient yields—often quite low yields but still sufficient—in bad years. Good yields can be achieved in bad years with more technology—notably, by irrigation. Irrigation can be good, of course: the annual flooding of the Nile that features so graphically in the Old Testament was an infinitely renewable benison that provided traditional Egypt with one of the finest agricultural systems the world has seen, until the Aswan Dam put a stop to it in the 1950s.

Irrigation is immensely valuable. One third of cultivated land is irrigated, and that land provides two-thirds of the world's food. But irrigation carries many dangers: from soil pollution to the creation of deserts in the places whence the water is drawn. If it is overdone, then the bonanza quickly gives way to desert. Huge tracts of western and southern Australia have already been lost to salt, dragged up from below. The logic of the present global market says "So what?" In the short term there are fortunes to be made, and the future can take care of itself. Technology can always find a way, we are told. But it doesn't always, and this is no way to run a world.

Even more broadly: very high yields are commonly achieved not simply by new and fancy crops, not simply by irrigation, but by huge inputs of industrial chemicals: fertiliser (including N), pesticides, fungicides, herbicides. In "conventional", meaning industrialized systems, the soil itself is generally taken for granted, with no specific attempt to retain its texture or volume. (In absolute contrast, in organic systems—at least as conceived by Lady Eve Balfour, who founded Britain's Soil Association—the soil is given priority.) Very forgiving, deep soils, with a perfect balance of clay (holds water), sand (drainage) and organic material (texture, fertility) in temperate climates (tropical soils don't readily retain organic material) can withstand the rigors of intensive exploitation for decade after decade. Some even improve over time. After all, heavily fertilized soils of particularly favorable structure and in sympathetic, temperate climates can increase their organic content year by year because a large proportion of the total biomass of any crop grows underground in the form of roots which remain after harvesting. But most tropical soils, and many temperate soils, are soon eroded by such bullish treatment. So it is that the wonderfully fertile soils of Britain's East Anglia, our cereal belt, have sunk by several metres since medieval times. They are easy to plough because they are rich in peat, the pickled remains of ancient moss; and as the soil is turned the peat is exposed and oxidizes away. The cereal fields of East Anglia must be actively drained, by pumping. But the pumping becomes harder as the soil shrinks and dries and blows away and soon we will be down to bedrock clay, and the fields will not be worth draining at all. Then the area will flood; and, because of global warming and rising sea-levels, the flooding will be quicker and more extensive.

The powers-that-be of confident (though now rather anxious), trading Britain, clearly feel that it is acceptable to write off our principal source of home-grown wheat and barley. The same kind of scenario—different in detail but similar in principle—is being repeated a thousand or a million times worldwide. The rapid conversion of the Amazon rainforest first into poor grazing and then into desert, is the best-known example—but the devastation of the Cerrado, Brazil's dry forest, is even more dramatic. The gratuitous ploughing of Greek hillside and English downland is smaller in scale but equally horrible in its way. Overall, the destruction is a serious blow to the world as a whole and to all humanity. It results entirely from the desire to maximise yields in the short term, without regard for the long term. Without regard for our children, that is.

Clearly, the erosion of East Anglia, and eroded hillsides worldwide began long before the global market came on line. Even Homer complained of soil erosion, some centuries before Plato. Clearly, though, the current zeal for maximization of yield, made possible in the short term by heavy engineering and high tech, and urged on ever more shrilly by the desire and assumed desirability of maximizing turnover in order to maximise wealth, will make things a great deal worse, far more rapidly than was ever possible in the past. One of the most horrible aspects of the whole sorry scene is that we could be using our present science and engineering truly to create a safe world that is good for everyone. Instead, we are deploying science and the high technologies it gives rise to primarily and often exclusively for short-term profit. The fabulous insights that science provides us with, which could and should be among our greatest assets, instead are boosting a world economy and order that are threatening to kill us all off. It is as if the aim was to bring on the apocalypse. It is beyond belief. But it is the way things are.

What's wrong with adding value?

Adding value is good—if value is truly added, and is not otherwise harmful. I will pay a good price for a good loaf. Good bakers add value, and are among the world's most valuable citizenry (though of course, in this modern age, they have largely been replaced by vastly inferior high-tech mass production). I will pay a small fortune for excellent sausages, lovingly fashioned by people who know what they are doing. Good bakers and sausage-makers truly add value. They deserve our thanks. In a world that took serious things seriously, they would be the ones we honoured.

But value-adding in our modern, debased, simplified economy is something quite different. The core idea is not to add worth, but to increase expense: to increase the money that can be charged. Much of what is done is obviously spurious: excess packaging; French beans jetted with subsidized fuel from Kenyan hillsides to suburban supermarkets (only people who can't cook would dream of serving Kenyan French beans); bright red strawberries whisked for no worthwhile reason at all between continents, to be served out of season, as if their chilly wateriness was somehow life-enhancing. It is all so obviously fatuous I will not waste words on it.

What does require comment, and is usually overlooked, is that the entire modern livestock industry has almost nothing to do with good agriculture, or with good nutrition or gastronomy, but is entirely an exercise in value-adding. Again the industry is designed primarily to make the people who control it (usually not the farmers) enormously rich, and in this it succeeds spectacularly. But is horribly cruel to animals and people alike, creates horrendous pollution, and in general is a huge threat to the security of the Earth as a whole and hence to all humanity.

The nonsense of modern livestock

As outlined in chapter 3, livestock can enhance the biological efficiency of agriculture, and improve our nutritional status and gastronomic delectation—but only if the animals are produced in ways that reflect their own biology and the physical restraints of landscape. Committed herbivores such as cattle and sheep should primarily be fed on grass or browse, which is to say on leaves, which in biochemical terms means cellulose, which human beings cannot eat, and which are grown in places where it is hard to grow cereals. The omnivores, mainly pigs and poultry, should be fed as was traditional on leftovers and substandard or surplus crops. Both the herbivores and the omnivores may be given some grain to help them through the winter (so that there are more of them in spring, to make better use of the summer grazing) and to help cows through their lactation. But grain is given very much as a supplement.

However, farmers in principle face one huge economic hurdle which they have often found hard to overcome. The title of my book is in some ways too true. Feeding people can be too easy. If people ate as recommended in chapter 3 and as most of humanity has done through most of history and prehistory—plenty of plants, not much meat, and maximum variety—then we could well be fed on a fairly modest output. But this would mean that the market for food would be "inelastic". The demand could soon be met. As societies grow richer, makers of cars can grow richer in parallel, because however wealthy people may become they can always spend more on cars—one for each day of the week; an SUV and a run-about; a Lamborghini and a couple of Ferraris for fun; and so on. But if those same people are content simply to eat good food (plenty of plants, not much meat, and maximum variety) then the farmers miss out. Adult humans generally need about 1500 kcals per day just to stay alive. But even the richest cannot get through much more than about 3000, unless they take up rowing, or spend their evenings in the gym, or are prepared to grow fat. Where's the profit in that?

Meat solves the problem. If wheat is turned into bread, leavened or unleavened (as in chapattis and parathas and goodness knows what), which is the sensible thing to do, then we can all eat well for pennies (if we have good bakers). But far more cash can be generated overall if that same wheat is first fed to pigs and poultry—or indeed, in these extraordinary days when sound biology has been put on hold, to cattle. So it is, just to repeat the point, that half the wheat now grown in the world is fed to livestock, 80 percent of the maize, and well over 90 percent of the soya. The excess meat thus produced is mostly consumed in the west, where the main health problem is obesity and its related ills, brought about largely by too much animal fat, but what the hell. The extra production has nothing at all to do with sound nutrition or with fine gastronomy, because the cooking that goes with all this surplus livestock—the 99-cent burger and wall-to-wall fried chicken—is unspeakable even though, alas, it is all that many people have access to. Livestock fed on grain that we could eat ourselves does not enhance overall efficiency. Such livestock competes with us. By 2050, if livestock production continues to increase at its present rate, then the world's farm animals will be consuming the equivalent of four billion people—equal to the total world population in the 1970s, when many feared there were too many to feed. So by 2050 livestock will be increasing the effective population that needs to be fed by 50 percent: nine billion people, plus four billion animals. If present economic principles continue, then we can be fairly sure that if poor people in Third World countries are competing for food with western cattle, then the cattle will win. Just as they do now. We can pay more for cereal for cattle-feed than poor people can pay to feed their children. That's the market.

But livestock fed on cereal and soya solves the farmers' most fundamental economic problem. It removes the theoretical ceiling on production and hence on wealth. Ten metric tonnes of wheat would solve the nutritional problems of about 50 people for a year—if they ate a traditional diet. But if the same wheat is first fed to livestock, then the resultant meat would feed only about five people. So cereal production can be increased. To be sure, the meat market can be saturated too, so although the ceiling on production is raised by feeding cereal en masse to livestock, it does not seem at first sight as if the market has been made truly elastic. But there is an easy way around this. Traditional cuisines make use of all parts of the animal: tripes, kidneys, feet: bring 'em on. But although I have bought tripe in Spanish supermarkets, I have never seen it in an English supermarket. We British are already too rich for that. In practice, the livestock market can be expanded indefinitely because when there are enough animals to feed everybody, the traders can simply throw most of each animal away—all but the most expensive cuts, the chops and steaks. Then if you ask for tripe or kidneys they can proffer their standard mendacious excuse: "We don't stock them because there is no demand". Since, to an ever-increasing extent, thanks to the idleness and pusillanimity of modern, corporate-controlled governments, the supermarkets now control the flow of information as well as the food supply itself, they can ensure that the cheap cuts that have launched some of the world's finest recipes are written out of the act. A whole generation has grown up that does not know of their existence (and if told of it, pulls a face. They have been brought up to prefer additive-rich reconstituted chicken shaped into dinosaurs. Value-adding writ large).

In short, the production of livestock by modern means—turning cheap cereal and pulses into expensive meat—is the most flagrant and widespread exercise in value-adding of all. It is, indeed, extremely lucrative. It also ensures that humanity's chances of surviving in a tolerable world are hugely reduced. For good, modern industrial livestock production is extraordinarily cruel to livestock and producers alike, and of course is helping to eliminate our fellow species (by increasing the demands that humanity as a whole makes upon the Earth).

But the third requirement of standard business practice—cut costs to the bone, and then cut them again—is the most pernicious of all.

Cutting costs

Nothing has harmed our world and the people in it more than the frenetic desire, and the perceived need, to farm on the cheap.

For one thing, cut-price agriculture is extremely dangerous. To begin with, again as outlined in chapter 3, cheap farming is simplified farming— the biggest possible fields with the biggest possible machines to achieve economy of scale; single kinds of crops grown horizon to horizon—large-scale monoculture—so that all can be harvested together, with one sweep of the combine. Hedges, copses, and islands of higher ground among the paddy fields for horticulture, are swept aside. Biodiversity is lost and so too is the essential nutritional variety. In addition, because the vast crops are uniform, all are vulnerable to the same diseases. One particularly virulent strain of one pathogen could wipe out the whole lot.

It is particularly dangerous—and cruel—to produce livestock on the cheap. Britain's recent epidemic of bovine spongiform encephalopathy, BSE, was caused entirely by the perceived need to cut costs. BSE is a horrible disease that attacks the nervous system, first causing the animal to stagger, and quickly and virtually invariably leading to death. It is caused by a prion, the same class of agent (not really an organism; just a deformed protein) that causes scrapie in sheep and kuru in human beings. It is spread when animals directly consume tissue, particularly nervous tissue, of other animals. Normally prions spread only between animals of the same species. Kuru seems to be confined to people who have practiced cannibalism.

Until BSE turned up, cattle were not known to be susceptible to prion diseases. They do not generally eat flesh of any kind, and certainly not that of other cattle (though they may be exposed to placentae if they all give birth together, so there may be some flesh-to-flesh transmission). But dairy cattle are traditionally given extra feed, including extra protein—so-called "concentrate"—during lactation. Now that yields must be maximized, to maximise profit, they are given more and more concentrate. Cereals are the traditional protein source. Soya is now the most favored, which is why Amazonia and the Cerrado are being blitzed. But it can be cheapest of all to make protein concentrate from bits of dead animals, including dead cows. Whatever is cheapest must, of course be done. So in recent decades, cows have been turned into cannibals. Not nice; but that's modern business.

All went well (apart from the disgustingness of it) until the 1980s when restrictions were lifted on the method of preparation, to save fuel, and hence bring the cost down even more. Then it was found that cattle flesh can contain rogue prions (which may or may not have been acquired from sheep). They spread like wildfire. They also spread to humans who had eaten cattle flesh—in the form of vCJD (variant Creutzfeld-Jacob) disease, related to kuru. Variant CJD kills people just as surely and horribly as BSE kills cattle. A lot of people eat beef, and professor Roy Anderson, a leading British epidemiologist, at one point estimated that as many as 100,000 people might be infected with the prion, and almost all of them, from what we know of the disease, would die. His worst prognostications have not come about but the game is not over yet—CJD can have a very long incubation period—and several have died already.

Traditional farmers know that the prime rule of good animal husbandry is to interrupt chains of infection, and preferably to avoid them all together. But in the case of BSE, a chain of infection was actively created where none had ever existed before—for cows in a state of nature do not eat other cows. Brilliant, eh? But that's business. Nobody was blamed for this fiasco, of course, although a few civil servants, who of course were not named and in any case were not in any worthwhile sense guilty, were rapped over the knuckles in an extremely expensive official report that was not, of course, briefed to look at the overall farming strategy. Whatever way you look at it, the BSE-vCJD episode has been a disgrace: thoroughly bad practice that should have been condemned on grounds of common sense and common decency, elevated to official policy; and then covered up. But that's modern business—and modern government.

Britain's other great gift to modern agriculture came in 2001: the biggest epidemic in history of foot and mouth disease (FMD). FMD is caused by a virus. It affects all animals with cloven hooves such as cattle, sheep, goats, and pigs. It is highly contagious and spreads by various means, both through direct contact and through the air. It does not normally kill a healthy animal but it is unpleasant and seriously debilitating and if left to run its course may reduce the subsequent performance of the animal throughout its life: for example depressing milk yield.

FMD can be controlled by vaccination, though not absolutely: vaccination may merely mask latent infection. So it is reasonable at least in an island like Britain to keep it at bay simply by ensuring that the virus never gets into the country at all. This is how we controlled rabies through most of the twentieth century, and with great success. But Britain took rabies seriously. Rabies affects pets and people. FMD merely affects farm animals. So whereas it was extremely difficult to smuggle a dog into Britain through most of the twentieth century (and smuggling carried huge penalties) it was a breeze to bring in raw meat, which could be infected. Journalists demonstrated this for a stunt as the FMD epidemic ran its course.

So it seems that FMD was brought into Britain some time before 2001—although for all anyone knows the virus might have been lingering for years in sub-clinical form in upland sheep. We don't have enough vets these days to check such things (too expensive). But the

FMD virus certainly did arrive by one route or another and because Britain's livestock are not vaccinated against it, the virus found very rich pickings. The previous British outbreak of FMD was in late 1967. But although that outbreak was then the world's biggest it remained largely confined to the north-west, around Cheshire, which is a traditional center of dairy farming. In today's cut-price world local abattoirs have been closed and animals are whisked from one end of the country to the other for the dubious privilege of being killed in an EU-approved super-duper slaughter-house. Animals sometimes travelled long distances in traditional systems but traditionally they walked, and the journey took weeks, so any illness that any of them had became apparent en route. Nowadays the longest journeys are completed well within the incubation period even of a virulent virus like FMD.

So it was that the first cases in the 2001 outbreak were reported on February 20 from an abattoir in Essex in the south-east of England, but were thought to have come from a farm in Northumberland, nearly 300 miles to the north. Within days it had reached Devon in the south-west and the last reported case, on September 30, was from Cumbria, in the north-west. In effect it was chauffeured, tourist class, to the far corners of the country, and then for good measure spread to Holland and France. By the time it had run its course it had infected animals on 2030 UK farms, all over the country. In the absence of vaccination, control was attempted and eventually achieved by slaughtering infected animals and those around them. Four million were killed, and were burned in huge pyres that featured nightly on the television news (although politicians insisted that they were "exaggerated". Apparently the 100-foot-high flames were "got up by the press"). The overall cost was put at £10 billion, about one percent of Britain's GDP. Even the Royal Shakespeare Company was hit because Americans in particular very reasonably decided to give Britain a miss while the epidemic was on.

The FMD virus does not infect human beings—or not so far, anyway. But most human infections began as zoonoses (animal diseases) and we should always be alert to possible mutations, and the horrible death of so many animals is far from trivial. Besides, some people did die. Some farmers committed suicide—even more than has become usual.

Britain's politicians and the industrialists alike exhort us with vehemence worthy of John Wesley to practice obsessive hygiene.

Manufacturers of disinfectant are given carte blanche to warn young mums that unless their kitchens gleam like intensive care units their babies will surely succumb to something vile (although they present no evidence for this). Little old ladies who make cakes for the village fête are told not to break the eggs on the side of the mixing bowl, for fear of some shell-borne microbe. The officially appointed Food Standards Agency with great pomp occasionally advises the withdrawal of some additive or other, from among the many hundreds that lace the modern diet. Yet Britain's whole agricultural system is run on a wing and a prayer. If modern livestock production had been designed by a crack team of pathogens, they could scarcely have done the job better.

Yet there is worse. Most destructive by far is that cut-price agriculture puts people out of work. Indeed, in the interests of reducing costs, it is expressly designed to do so. This is discussed in chapter 5: it is so important it requires us, humanity, to re-think the way we are structuring the entire world. To anticipate: if all the world followed the western way of farming, as they are exhorted and often obliged to do, up to two billion people worldwide would be without livelihood; and despite the enthusiasm shown in high places for IT and tourism, and hairdressing and the re-cycling of old tires, there is nothing useful or sustainable for most of those dispossessed people to do. Unemployment is the royal road to poverty. Politicians and captains of industry assure us that they desire above all to "make poverty history" and in this they are supported by celebs of all kinds, to add street-cred. But the methods they advocate—more westernization: high tech and the global market—are guaranteed to create poverty on a scale that even in this miserable world is still hard to conceive. Even in this age of official absurdity the muddleheadedness beggars belief.

In short, although capitalism in some form or other may still offer the world its best options—it really should be possible to reconcile personal freedom with social justice, as the US founders envisaged—the present, simplified, consciously amoral form of it is highly destructive. At least, the ruthless global market may benefit some industries; but it is as antipathetic to the cause of enlightened agriculture—agriculture that is intended actually to feed people—as can be conceived.

Yet even people who can see the horrors and absurdity of the present-day food supply chain are apt to make excuses for it. Doesn't it provide us with cheap food—at least in the west? Wouldn't food be far more expensive if agriculture was run along more enlightened lines? In an age in which routine mendacity is merely a tactic, this is still among the most pernicious of all lies.

The illusion of cheap food

I shop when I can at a couple of traditional markets where beans are sold *en masse* in paper sacks and vegetables come straight out of boxes and it all costs about a third, and sometimes only a tenth, of what the same things would cost in supermarkets. Food is as dear as it is largely because of middle-men's mark-ups. Take away the fripperies, and the cost can come down dramatically.

Nonetheless, at least in the short term, food produced by the methods of Enlightened Agriculture would probably be more expensive than a lot of the food in modern-day supermarkets appears to be. It is impossible at this stage to say how much dearer it would be, for reasons that will become apparent in the following paragraphs. But, for example, a local farmer friend of mine tells me that his free-range chickens, raised under trees (chickens are basically jungle fowl) and fed a natural diet, take twice as long to reach market weight as those raised intensively in sheds, and require much more space, and so they are bound to cost at least three times as much as the "conventional" industrial kind. So the defenders of the *status quo* do seem to have a point. But actually, it's not much of one.

For in truth, *there is no such thing as cheap food.* If chickens ever sell in the supermarket for 75 cents per pound as they often do, or cans of fruit are offered at three for the price of two, then we can be sure that some person or society or animal or landscape, somewhere along the supply chain, is being screwed. Some farmer is working for less than the cost of production; his workers are paid slave-wages; the animals are packed in cages, with the lights dimmed, and a body-full of growth-promoters; some hillside is being eroded, some forest felled, some river polluted—and all the creatures who used to live in those hills and forests and rivers, and all the people who enjoyed them and made their living from them, are being swept aside. Or, as Britain's epidemics of BSE and

FMD demonstrated, farming is being run on a wing and a prayer. Or, as often is the case when supermarkets seem sometimes to give the goods away, some market ploy is afoot, destined to put some local trader out of work. The food is cheap only because, for various reasons, the true costs are not taken into account. Who picks up the bill for the local fishermen when a prawn farm wipes out the mangroves, where the fish breed? Who cares? But somebody, somewhere, is suffering; and sooner or later, all of us will be picking up the bill.

Overall, too, supermarket food is not intended to be cheap. Individual items may be cheap, to be sure, but a key component of the market economy is to "add value"; which means, to make dear. Traditional cooking based primarily on fresh, local ingredients should in principle be able to supply far better nutrition and gastronomy more cheaply. If it currently fails to do so it is because the present economy is rigged in favor of mass production and mass long-distance transport. Food from traditional markets is not gratuitously packaged and processed. Traditional people do their own cooking. Even more to the point, traditional cooking exemplifies the general rule both of sound nutrition and of excellent cuisine: plenty of plants, not much meat, and maximum variety. The industrialized suppliers, by contrast, strive to push us ever further up the food chain—since meat is innately more profitable than staples. You can still eat wonderfully for pennies in village Turkey not simply because of the exchange rate but because the food is variations on a theme of wheat and honey and herbs, raised locally, and cooked in the back of the shop by people who know what they are doing and do not need to waste money on marketing and neon signs. One of Britain's biggest supermarket chains recently sold full-sized (what used to be called "two-pound") loaves for 20p. Their television ads in effect acknowledged that the bread was rubbish. It was best toasted, they said (although lousy bread makes lousy toast). Yet in truth there is nothing dearer than a 20p loaf. If bread is good, you can eat a great deal of it. When France was at its culinary height the French ate bread at every meal, and often with very little else; just a touch of cheese and sausage and the occasional leek. But what bread! I remember how magnificent it was. Such bread can still be found, here and there. The memory of it is not an illusion. If bread makes up half your diet, then it is cheap at $2.00 a loaf. Uneatable bread merely provides wrapping for whatever is considered more esculent but relatively speaking will be extremely expensive (like industrial chicken in industrial mayonnaise with a token frond of damp lettuce. As Snoopy was wont to say: "Eeaagh!").

But let us shift the argument. *Why* should food be cheap? Again, half a century ago when their food was still great, French people on average spent 30 percent of their income on food. Why not? They built their lives around it—and very civilised they were too. You didn't have to be middle class, and certainly not a poseur, to be a gourmet. Some of the finest cafés of all time were designed for French truck drivers. Even now on the continent of Europe—France, Italy, Spain, Greece, even Germany—you see entire extended families gathered around vast restaurant tables in the evening, with the children falling asleep *in situ* ("It takes a village to bring up a child," as they say in Africa); not rich families, but ordinary working people. What is better? What else is life for, if not for such sociality? The British at present spend only eight percent of their income on food—and still "demand" that it should be cheaper. Why? Perhaps because so much of it is rubbish, and the less you spend on it, the better. If it was good, perhaps we would see the point of spending more.

Finally, though, we should question the sanctimonious argument that food must be cheap for the sake of the poor. In countries that have always been poor, food has often become far dearer in recent years because home-grown food has largely given way to commodity crops, sold for cash that stays in the hands of the entrepreneurs. In countries like Britain and the US, both fabulously rich by world standards, we should ask, "Why do they have poor people at all?" There are many reasons for poverty of course but the general answer, in both countries, is injustice. Many people in these rich countries remain poor because the economy is designed above all to ensure that some people are extremely rich. The economy is designed to be maximally competitive (although if you are rich enough you can cheat) but it is not designed to be equitable. The antidote to poverty is to give a damn, and create economies with fairer shares. The answer is not to be cruel to animals, or to screw farmers into the ground, or to fell forests and pollute rivers. These are the methods of the scoundrel.

In short, the globalized market does not serve us well. It may work well for motor cars and white goods but agriculture is different, and it is demonstrably bad for agriculture. Farmers are in the front-line, and they are suffering already. China, India, South Korea and others are apparently doing well out of the global market because they are making the things that the market is good for. But all their farmers are suffering—in South Korea the suicide rate has become horrendous—and as globalization bites they will suffer more; and farmers and their families, worldwide, account for about four-tenths of all humanity. Soon we will all suffer—and indeed are already suffering—because agriculture that really works and is sustainable needs good farmers, and plenty of them. It needs to be labor intensive. All of us need it to be labor intensive. The economic system that is making it impossible to be labor intensive is killing all of us.

So what can we do about it?

Can we do better?

The short answer, I believe, is yes. Various think-tanks and individuals around the world are on the case, taking their lead both from ecologists and from professional economists. They include Britain's New Economics Foundation, cofounded by James Robertson and his wife Alison Pritchard. I recommend the NEC's and James Robertson's websites. Many practical schemes are already up and running. These include cooperatives of many kinds; trusts; clubs; social franchise; micro-credit; local currencies. If these initiatives and models can be built upon, and if they can be coordinated, then already, perhaps, we have the basis of what the world needs.

In short, the task before us is not to confront big governments and the corporates, for that is merely exhausting. We need instead to create viable and clearly superior alternatives, and allow the status quo to wither on the vine. To do this, we need only to build on what exists already. In the last chapter I discuss how we might bring the necessary change about within the specific and crucial context of food. In the next, penultimate chapter I want to broaden the discussion and ask what a society would look like if it really did have an economy geared to its own wellbeing, and really was concerned for its own children and for the world as a whole. Such a society would, in fact, be agrarian.

Chapter 5

THE NEW AGRARIANISM

Sometimes the word "progress" has spiritual and personal connotations, as in John Bunyan's late seventeenth-century *Pilgrim's Progress*. Sometimes in the Whiggish tradition it is applied to societies—a steady increase in tolerance, justice, and literacy. In the main, though, in these secular days, "progress" is taken to mean industrialization—the replacement of human labor and of craft with machines and factories; organization— the government knows where you live, and so can ensure that you pay tax; urbanization, for cities are much tidier than higgledy-piggledy villages and farms; and increase in wealth, measured as "Gross Domestic Product", or GDP. Economic growth—rise in GDP—has become the prime index not simply of "progress" but of "development": an even grander concept.

Traditional farms do not make as much money as factories and call-centers and sweatshops may do and so they must be swept aside. The agricultural work-force is being replaced as quickly as can be arranged by heavy machinery, industrial chemistry and, now, biotech. The rural communities are packed off to the cities, their houses sometimes taken over by urban commuters, and sometimes left to fall down. Indeed, agrarian societies—built around labor-intensive farming—have become the world symbol of backwardness. Small farmers worldwide are perceived as an encumbrance, almost as a disease: a strain on the economy; an insult to the modern ideal of the designer tee-shirt and the cappuccino. They are also perceived to be unhappy—suicide worldwide has reached epidemic proportions. So it is a favor to throw them and their families off their land. To suggest, in this self-assured and well-heeled world, that the world's economies must again be basically agrarian, seems ludicrous.

Yet this is precisely what I am suggesting. For reasons that to me seem inescapable, human beings have very little chance of getting through this present century and into the time beyond *unless* our overall economy remains agrarian. The task for humanity is not to sweep the agrarian way of life aside but to make it work. This is not a recommendation simply to return to the past, however; the future agrarianism must be traditional in overall structure, but often very different in detail. We need to design "The New Agrarianism", and to make it the norm.

Why must the world be agrarian?

We need the new agrarianism for four main reasons.

First, the world needs agriculture of the enlightened kind—designed expressly to feed people, and to look after the fabric of the world; and since enlightened agriculture requires skill and is complicated, it needs a lot of hands on deck. Contrariwise—the second reason to be agrarian—low-labor farming is extremely risky. Thirdly, farming is still the world's biggest employer by far—and if we look at the world's resources realistically, we see immediately that it always must be. Indeed if we produced all our food in factories (as in the 1970s some suggested we should) it would still be a good idea to practice farming, just to provide interesting work. Finally, if people don't live in the countryside they have to live in the cities—and the cities are already overwhelmed.

The first point—that we need enlightened agriculture and enlightened agriculture needs a lot of skilled people—was discussed in the last chapter. So what of the second point, that low-labor farming is dangerous?

To begin with, just as labor-intensive farming can be as intricate as we want it to be, so low-labor farming must be simplified. Modern farms are as big as possible and are typically devoted to single crops—the approach known as "monoculture"—so that the land can be prepared and the crops harvested by the biggest possible machines. Thus is achieved "economy of scale". So we see maize (corn) in North America's former prairies, and wheat in Britain's eastern counties, stretching to the horizon.

But monocultures are extremely vulnerable. All the individual plants are susceptible to the same strains of disease and an infection that can attack any one of them is potentially able to wipe out the whole lot. So the crops need or are perceived to need extra quantities of herbicide, fungicide, and pesticide; and since labor is cut to the bone there is no time to assess the crops to see if particular pests are in practice causing problems and so the toxic brew must be applied prophylactically, in anticipation of possible outbreak, just by following the manufacturers' instructions. Proper farmers call this "farming by numbers". Agriculture becomes an exercise in industrial chemistry, abetted increasingly by genetic engineering to make crops that are better able to withstand the chemical assaults. Meanwhile the copses and hedges that wildlife need and the patches of horticulture which in the Third World in particular people depend upon both for gastronomic variety and for micronutrients, are rooted out. They get in the way of the machines. The land is deemed too valuable to waste on human communities and ways of life, or on the survival of wildlife. There are profits to be made.

Extinctions result, and nutritional deficiencies. Among the more conspicuous these days is hypovitaminosis A—lack of vitamin A, leading to "dry eye", alias xerophthalmia, which the WHO says is currently causing blindness in up to 500,000 children per year worldwide. Vitamin A (or at least the precursor of it) is carotene, the yellow pigment in plants. Carotene is an extremely common molecule. It occurs in quantity in dark green leaves such as spinach, in yellow roots such as carrots, and in yellow fruits such as papaya and mango, which grow like weeds throughout the tropics, given half a chance. The sensible long-term antidote to xerophthalmia is to restore the horticulture that has been swept aside by industrial monoculture. Instead the genetic engineering industry has produced "golden rice"—rice that contains carotene genes—to ameliorate (at least up to a point) the effects of its own monocultural excesses. Thus it compounds its own nonsense. Again the experts emerge as *idiots savants*, focused exclusively on their own small skills and trades, utterly ignorant of context (or else exceedingly cynical). But the creators of golden rice will probably be rewarded with Nobel prizes.

Of course, too, farming that is maximally industrialized uses enormous amounts of fuel; and since masses of any one kind of food are produced all in one place it then has to be shipped, and sometimes jumbo-jetted, around the world. All this while the oil-wells are running dry and the atmosphere is turned into a greenhouse by surplus CO_2. Monocultural farming also lacks versatility—a few crops with the same genes grown on the largest possible scale. Yet in times of climate change versatility is a prime requirement. A six-year-old could work that out. But not the powers-that-be, fixated on whatever is cheapest, *now*.

But it is with livestock that the dangers of low-labor farming are most evident. Traditionally, a farmer might keep a dozen dairy cows as one component of a mixed farm. He or she knew each animal like a member of the family: when it was born, who its parents were, its illnesses, its temperament, its likes and dislikes. Farmers had plenty of time to establish such personal relationships: typically, dairy cows on traditional farms lived a dozen years or so, producing 10 calves along the way, each birth followed by a lactation in which each cow produced, typically, 2500-4000 liters of milk (500 to 800 gallons) over 10 months, which is two or three times as much as a wild cow would produce for her own calf.

In modern, industrial dairy units one worker may "manage" 120 cows or more, each of which is typically expected to produce at least 5000 liters in each 10-month lactation. Indeed the 10,000 liter (2000 gallon) Holstein is already commonplace—producing about six to eight times as much as a wild cow. But dairy scientists already dream of the 20,000 liter (4000 gallon) cow, who would need to be milked more or less continuously (I have heard them dreaming such dreams at conferences. Implants of hormones and genetic engineering are the perceived routes.).

Traditional dairy cows were lovely animals, and still turn up in shows that are put on, as show-biz, for public consumption: Jerseys, Ayrshires, Shorthorns, dual-purpose Devons, and so forth. But modern industrial dairy cows, as featured in commercial brochures and trade magazines, are hideous: huge bags of milk on minimal skeletons. These benighted animals are lucky to survive two lactations, let alone 10; 1.8 lactations per cow is the current British average. But although the modern cows and the machines that milk them and the drugs that keep them roughly on their feet cost a fortune, the overall system is "economic" because there are so few employees and the output is so enormous and besides, in the European Union and the US, such systems are subsidized with taxpayers' money. We taxpayers-qua-consumers are rewarded in turn, or so the policy makers deem to be the case, with milk that sells in supermarkets for less than bottled water. Also, some of us have shares in the companies that are making a fortune from such exercises, or our pension schemes do.

In the United States, multi storied piggeries with a million animals are now almost commonplace and, again supported by taxpayers' money, they are planned for Europe; not least for Poland, which traditionally employed many thousands of people raising pigs in traditional ways. The cruelty of such units both to livestock and employees beggars belief while the pollution problems are horrendous. One pig produces as much ordure as eight people. A million-head piggery produces as much sewage as London. On traditional farms, manure is a valued asset. In mega-units, with no surrounding crops, it is a nightmare. But then, as outlined in the last chapter, the companies who create the nightmare can bend the law to absolve themselves blame, and carry on as normal, if they are big enough. Again as outlined in the last chapter, low-labor husbandry is cut-price husbandry, and that is immensely dangerous. The epidemics of BSE and foot-and-mouth disease (FMD) that were Britain's contribution to world farming during the last decades of the twentieth century, resulted entirely from the perceived need to cut costs, and then to cut costs again.

Most important, though, is the third reason for keeping people on the land. It employs people. Follow the path of the modern progressives and we will put nearly half the world's workforce out of a job. Unemployment is the royal road to poverty. Poverty is what we are supposed to be at war with, if you believe presidents, prime ministers, and organizers of pop concerts. But the more we contrive to industrialize and monetize the world's farming in the spurious cause of "modernization", the more we create poverty. "Modernization" as currently perceived means the end of small-scale farming. But for most of the world's two billion or so traditional farmers and their families, there is nothing else useful for them to do.

How many farmers is enough?

Of course, a country can have too many farmers. In Rwanda and Uganda 90 percent of the workforce works on the land. In Angola it is 80 percent. We can all agree that this is too many. The Rwandans and Ugandans agree with this too: this is not high-handed European imperialist talk. The 10 or 20 percent who are not on the land are not enough to fill all the other jobs that societies need: teachers, builders, engineers, doctors, nurses, cooks, shopkeepers, scientists, priests, theologians, philosophers, historians, artists, musicians, writers, cleaners, athletes, accountants, lawyers and administrators (every society needs a few, provided they know their place). A society that contains nothing but farmers is at least as bad for the farmers as for everyone else. It condemns them to a life of toil and nothing else, and their descendants to the same. Farmers need other people doing other things both to supply them with a market and to supply life's other needs and pleasures.

On the other hand, no one who is not steeped exclusively in money, and whose life is not lived solely between the bottom line and the right-hand column, and who knows anything at all about life's biological or social realities, or who reads the newspapers, can possibly suppose that it is good to employ only one percent of people on the land full time, as has become the case in Britain and the US. You have to be very foolish indeed to believe that—and of course the ideal of minimal labor is a fake, because both in Britain and the US the farm workforce is propped up with immigrants of conveniently dubious legal status, with conveniently ambiguous human rights. It is a symptom of how far the world has sunk, of the poverty of understanding and feeling among those who determine the course of the world, that Britain and the US, who in truth are so far out on a limb, are held up as models for the rest of the world to follow.

If 90 percent is too many, and one percent is too few, what would be the ideal proportion of people to work the land in any one country? Since two billion people worldwide are working on farms—about one in three of all of us—this ought to be the subject of urgent debate. Every other social issue seems dwarfed by it. But of course it is not. The powers-that-be simply take to be self-evident that whatever course is cheapest, in the immediate term, and returns most profit to the world's most powerful companies, is what we must do. Everything else is "unrealistic". In general, one of the most alarming features of the modern world and of the powers-that-be who run it is that the really important questions are not discussed—and indeed are not even conceived.

Since the powers-that-be have yet again demonstrated their lack of grasp, let's begin the debate right now. First we might reasonably ask, if people are thrown off the land in the interests of "modernization", what else might they reasonably do? Well, clearly, there are a lot of them. In the Third World as a whole, 60 percent of the workforce are on the land. In India, the workforce is officially numbered at 480 million. Sixty percent are on the land (the Third World average) making 288 million.

This is roughly equal to the total population of the United States; but since the workers have dependents, we may reasonably guess that the total Indian population that is directly dependent on farm labor is around 600 million—far more than the total population of the newly expanded European Union.

In recent years the British government, with British commercial interests at heart as it has had in India for the past 400 years, has been encouraging Indian agriculture to industrialize: for "them" to be more like "us". If the Indians do follow our government's advice they will put at least 250 million people out of work.

What else can those 250 million do? I have asked this question in India. I was told (not by everyone; but by zealots for westernization) that there are "alternative industries". What industries? Ah—well, there is IT (information technology); and, er, tourism.

India has an IT industry of world class. Why not? It has more than three million science graduates (with a good sprinkling of Nobel prize-winners). Bangalore has become India's "silicon valley". But, according to the *Hindu Times*, the IT industry of Bangalore employs just 60,000 people, all of them graduates. So there is one IT job for every 2500 farmers (and very few traditional farmers are graduates). Official Indian estimates far more generously suggest that the total employed in IT defined broadly is now around one million. That is still less than half of one percent of the farmers who would be out of work if India's agriculture was industrialized.

Millions, doubtless, are employed in tourism. I have spoken at length to some of them, particularly taxi drivers in and around Delhi. There is plenty of time to talk, as you sit in the traffic. The drivers work an 80 hour week for about $12 a month. They will wait for you all day for a $1.50 fare. They see their families at best for an hour or two each day, in a state of exhaustion. In Brasilia, Brazil's show city, I failed to meet the many thousands employed as hotel cleaners: failed to meet them because they are bused in from the distant shanty suburbs before dawn, before the refined eyes of northerners can catch sight of them, and then bused out again. Farm-work can be hard, but why is this version of city life supposed to be better? But the taxi-drivers and the cleaners are the lucky ones. They at least have jobs.

To some extent, India's dispossessed farmers have taken matters into their own hands. Between 1993 and 2003, according to India's National Crime Records Bureau, 100,248 of them committed suicide (*New Internationalist*, September 2066, p 17). Nonetheless, I have heard British politicians and captains of industry (including at least two newly promoted to the British House of Lords), zealously extolling present trends. India and the Third World as a whole, they inform us, are re-enacting the history of Britain itself. Four hundred years ago Britain too was primarily agrarian—and look at us now! Almost everyone has a fridge and a television and can look forward to a state pension. "Poverty" is still an issue in Britain—the evidence that it is diminishing is highly equivocal—but on the whole, as Prime Minister Harold Macmillan said nearly half a century ago, "You've never had it so good". If you suggest that the Third World as a whole cannot follow the western lead, then the zealots for progress are liable to accuse you of racism. Do we dare to suggest that only Europeans can run factories?

But that, of course, is not the point. For argument's sake we will skate over the downsides of modern western life. Life is seriously unpleasant and dangerous in many a northern British town these days, especially if you're the wrong color, and on the whole I would rather not be in Philadelphia. In contrast, you can feel very at home and safe in Indian and Indonesian villages, surrounded by smiling faces; and although life was obviously hard on a tea plantation I stayed on recently in Kerala, it was clearly very jolly, too. The children play outside, unself-consciously. Neither are they "doomed" to stay on the plantation forever because they go to school and the bright ones move on to the universities, just as in any other well-run and humane society. Neither are they ferried to school as they are in my native south London in SUVs as heavy as armored cars, for fear of other SUVs and real or imagined pedophiles. Zealots for "progress" like to tell us that all references to traditional ways of life are mere nostalgia and therefore, apparently, are bad. But if you look with unprejudiced eyes it is plain that much of the world we are now creating is vile, and much of what was there before, was not. How can it be progress to trade what is agreeable for what is not?

Perhaps, though, this is just a matter of opinion. I have not actually lived as a native in an Indian village or in a Louisiana trailer park, so who am I to judge which is preferable? Less equivocal are simple matters of fact. Modern western life is tolerable if you are up to speed economically.

Not otherwise. But as everyone surely knows these days, it would take the resources of three planets Earth to raise everyone on Earth to the material standards of the average Brit, and five or more Earths to bring us all up to the average Californian; so the dream of western wealth for all is obviously a mirage. As Gandhi pointed out, Britain depended for its meteoric industrial rise on the wealth of its empire—the countries that now form much of the Third World. Who, asked Gandhi rhetorically, will be the Third World's empire?

Or then again, although the *proportion* of the workforce in British farming declined steadily as our industrial revolution gathered momentum after 1800, the actual *numbers* did not fall substantially until well into the twentieth century. Britain's growing industries, fed by the resources of the empire, sucked in the surplus workers from the land, and then from the growing cities, and finally from abroad as we still do—eking out our labor force with West Indians, Asians and, now, Africans and East Europeans. Only after Britain's factories and all that goes with them had been up and running for the better part of two centuries did Britain finally industrialize its agriculture, and remove the traditional farm workforce. If and when Rwanda or indeed India have industries that are truly able to employ the people who now work on the land, then will be the time to invite them into the cities. To destroy the traditional farms and the employment they provide *before* there are such industries—or indeed, when no such industries are even on the horizon—seems to me careless to the point of wickedness. But there never can be such industries on the scale required. The Earth simply isn't big enough to supply the necessary resources. In particular, the industries of the west have been fired by fossil fuels that have been superabundant and ridiculously cheap, and those days are past. Yet people are being urged with all possible vigor to industrialize their agriculture and leave the land. Again, the sheer lack of grasp among the people who have most influence beggars belief.

And where are the ex-farmers and their families supposed to live? In 2006, for the first time in the world's history, the number of people in cities equaled the number in the countryside. By 2050, on present trends, two-thirds will be in the cities. The population in cities will then equal that of the *whole* population of the present world. At least a dozen cities by 2050, on present trends, will have more than 50 million people each: each roughly equal to the population of present-day England.

2050 is not far away. I hope and expect that all of my children will still be around and my grandchildren should still have half their lives before them. Yet it is already clear that the cities cannot cope. Already, according to the United Nations, a billion worldwide (one in six of everyone) are living in urban slums. The Chinese Government, these days, is in many ways enlightened and I am reliably told that they would like to curtail the migration of 500 millions to their cities. But, they do not quite have the control that the outside world imagines they have. The flood continues. The priority there, as everywhere, is to make the countryside agreeable.

Overall the feeling grows in high places that urbanization is inevitable— not just a matter of conscious progress but a kind of Hegelian social evolution; and from this feeling has emerged a literature on the alleged pleasures of modern slum life. People adapt. In what seem the vilest conditions they make a living though a myriad of services one for another—dressing hair, fixing bikes, restoring old radios, making ingenious toys and household goods from tin cans, polyethylene bags, and—especially prized—old tires (and there is always prize-fighting, prostitution, drug-running, professional begging, and mugging). There is a new expression: "favela chic". Again, though, I wonder why a life spent rescuing debris from the garbage dump, among the dead dogs and the abandoned pizzas and the somewhat verminous kites, is meant to be preferred to life on the farm. More broadly: people in slums are demonstrating the wondrous ingenuity of humanity, and resilience of spirit. If this ingenuity and spirit could again be allowed to function within the world at large, what a world we might have! In the present world, though, dominated by engines for making money, mere human spirit has been banished to the slums.

So—what proportion of the workforce should work on the land? I do not presume to know. The necessary debate is not taking place. It is taken to be self-evident in high places that fewer is better—that indeed the world will not be fully civilized until all rural work is carried out by robots (which is almost technically possible now, provided the agriculture is kept very simple). But I suggest, as a working hypothesis, at least to set the debate in progress, that the rural workforce in any one country should probably not be greater than 50 percent, or less than 20 percent. The 50 percent maximum would leave plenty of scope for other trades and professions and opportunities, while ensuring (if the farmers were competent) that the food supply was as secure and agreeable as possible and the land stayed in good heart. This would mean that Rwanda, say, needs quite a lot of reform, but that India is probably on course already. By contrast, Britain and the US and the other rich western countries that are striving to catch them up, have gone far too far. They are in a mess themselves (although propped up temporarily by the surpluses and misfortunes of others) and are setting the worst possible example to the world as a whole. For them to urge others to follow in their wake is irresponsible in the extreme (although whether this is evil or merely misguided I will leave others to judge). Again we see that it is ourselves, not the "Third World", who are most out of touch with reality, and most in need of transformation. Yet the whole thrust of our economy and our political philosophy is to continue as we are, and to encourage others to follow.

Mere tweaking of policies is not what's called for. A whole new mind-set is needed. That is precisely what we are not going to get from conventional leaders. That is why those of us who give a damn just have to start doing things ourselves.

Much more would follow—all desirable—if we farmed properly (according to the tenets of enlightened agriculture) and established new agrarian societies. Most of us would find that most of our food was supplied locally, and therefore fresh (and we would know the farmers who produce it); and virtually all countries in the world (and certainly all the countries in Africa that are now considered most disastrous) would become self-reliant in food. Countries that were self-reliant, in turn, could begin to put their own trade into perspective. The world's poor countries (Africa again comes to mind) would be far less dependent on the good offices and charity of the West. The most powerful western governments, and the corporates who support them, wouldn't like it at all.

Local produce, self-reliance, and fair trade at last

Though there would still be some specialist plantations and ranches under a system of enlightened agriculture the typical farm worldwide would be small, often family-run, and mixed: arable, horticultural, and pastoral all integrated, with several or many kinds of crop or livestock in each. Tropical farms would have tropical crops—rice, maize, sweet potatoes, mung beans, cardamoms, cinnamon, mangoes—and temperate farms would have temperate crops—wheat, potatoes, broad beans, basil, apples, raspberries—though there would be much that was found everywhere: cattle, pigs, chickens, ducks, sheep, spinach, onions, and so on. The cuisine would vary region by region and season by season and *vive la différence*— but everywhere we would find the same, basic, unimprovable nutritional balance: plenty of plants, not much meat, and maximum variety. Because each class of crop and livestock would be grown everywhere (even though the species and varieties differed from place to place) everyone could find a balanced diet and the bases for local and brilliant cuisines within a few miles; or at least they could, if they were not packed into cities of 50 million-plus. Every nutritionist, cook, and foodie knows that fresh food is best. Only the supermarkets and the processors, who make their fortunes by ferrying French beans and watery strawberries across the world and the seasons would feel in any way incommoded.

If all countries grew all the food they need, on their mixed farms, then they would *ipso facto* become self-reliant in food. Self-reliant does not mean self-sufficient. Self-sufficient means growing absolutely everything for yourself, and eschewing trade all together. Some countries have tried doing this in the past (as China did for a time after its revolution) and some have had self-sufficiency forced upon them (as Britain more or less did during the two world wars) but for most countries this is not a desirable state. Britons famously like bananas, and although we could grow them ourselves in greenhouses, this would be an awful waste of fuel. And everyone likes tropical spices.

So of course, in a world rooted in new agrarianism, there should still be trade in food. But it would be run on common-sense lines. Intellectuals of all kinds, including scientists and economists and moral philosophers, feel they need to improve on common sense, for how else can they justify their existence? But in truth, common sense is very hard indeed to improve upon. Common sense says that it is good to trade for all kinds of reasons—you turn your own surpluses, the things you are especially good at growing, into cash; and you have access to nice things, like bananas and cardamoms, that you cannot easily grow yourself.

Common sense says too, though, that it is dangerous indeed to rely utterly on trade. At least, you can get away with it if you are, or were, say, Britain, in the nineteenth and early twentieth centuries, with a fairly complaisant empire and the boundless wealth that goes with it and warships to blast all opposition out of the water—although even Britain was horribly caught out in both world wars, when the trade routes were severed, and in particular the Atlantic was blockaded. But if you are, say, Ghana, or Costa Rica, in national debt to the rich countries whom you have helped to make rich, and with no serious sabres to rattle, then absolute reliance on trade is the royal road to perpetual subservience. Of course world trading is unfair. A few people call the shots, and most do not. Worst placed of all are countries such as Senegal and Costa Rica that stopped growing all their own food—which they used to do easily—so as to grow more ground-nuts or coffee or whatever for export. This of course is what the powers-that-be encourage them to do.

Poor countries are hopelessly trapped. The World Trade Organization which is supposed to provide a level playing field for all is in reality controlled by the richest governments which in turn are beholden to corporates. On the other hand, if poor countries flout the WTO they are liable to find themselves bound in bilateral agreements with the US, on terms that are even less favorable. So it is for example that modern-day Costa Rica no longer grows the maize and beans that are native to the country, and used to support the people, but instead buys them in from the US, where they are grown at several times the cost but are subsidized by the US taxpayers; and sold in Costa Rican shops for far more than the old native-grown crops. Meanwhile the western leaders shed crocodile tears for the world's poor and hungry and discuss very minor increases in aid, which they recover several times over in interest on spurious loans. This system *might* work if it were fundamentally benign; if, deep down, the powers-that-be actually wanted to help the poor. But they do not. Other people's absolute abjectness is an embarrassment because abject people cannot function, and they have no buying power. But relative poverty in others—just about rich enough to function and provide cheap labor but not rich enough to compete—is to the advantage of the truly rich.

The only way out for poor countries is autonomy: *not* to be beholden to rich countries who basically are not on their side, and deep down prefer them to be relatively poor. Outstanding among the countries in recent years that have tried to go it alone is Cuba which, in the 1960s, among other things, managed in short order to raise Charolais cattle for beef on sugar-cane when the US decided not to buy its sugar any more. Cuba perhaps is not the most comfortable country to live in but it does look after its poorest people in a way that many—most—westernized countries do not. The basic reason why Cuba can hold its head up, and Costa Rica cannot, and has nothing to look forward to but decades of compliance, is that Cuba grows enough food to get by on. When the chips are really down, as indeed they are, Cuba doesn't need to trade. If all countries in the world were in that position, and if all countries were free to trade only in the commodities they had in excess, and for which they had guaranteed markets, the world would look very different. All countries could trade when it was to their advantage, and not when it was not. No one would be forced, as now, to trade on terms that are so obviously self-destructive ("... another day older and deeper in debt", as Tennessee Ernie Ford put the matter). I suggest that in the absence of universal good will, which has never prevailed and in this hyper-competitive world is actively suppressed, self-reliance in food for all countries that lack the muscle-power of the west is a *sine qua non*. If countries like Ethiopia and Angola and virtually every country in Africa had new agrarian economies geared to their own local variety of enlightened agriculture then they could be self-reliant in food several times over.

In short: enlightened agriculture leads to the new agrarianism which provides the best and freshest possible food and leads to national self-reliance worldwide which at last would provide the only robust basis for fair trading. Abandon the basic principles on which enlightened agriculture is based—common sense, common humanity, sound husbandry, and good biology—and we are bound to sink deeper and deeper into the mire of hunger, dependence, and global profligacy. This is exactly what we are doing.

Hanging over all human endeavour, of course, and indeed all life on Earth, is the spectre of climate change.

What do we do as the world warms up?

Global warming will produce uncertainty everywhere. For Europe to become reliant on soya grown on the Brazilian Cerrado when the Cerrado itself might become desert, or on Canadian wheat when the wheat refuses to grow, would be disastrous. Even if there is food to buy on the global market, who will be able to out-bid China, whose own fields seem all too likely to fail? The Gobi is currently encroaching on Beijing at 30-50 km a year. I have seen the maize dying on the Beijing outskirts, but the farmers go on planting it even though they know it will die, simply because they are farmers and planting is what farmers do. Although almost all countries could be self-reliant as things are we cannot guarantee that *any* country will be able to grow enough food for itself in 50 years time. The safest course, so common sense tells us, is for everyone to try to grow all the food they need everywhere, in the hope and expectation that some at least will succeed and perhaps have surpluses for the rest—although we cannot at present say which will be in surplus and which will fail altogether. For a country like Britain deliberately to run down its agriculture in a time of such global uncertainty seems like madness; and to encourage poorer countries to cut down on food crops just to grow non-food crops for export, again seems very like wickedness. Present world food policies have drifted so far from common sense and common humanity and sound biological thinking as to leave us gasping. But the people who make these policies are extremely rich and they take this as sufficient sign that all is well with the world. The annual summary for 2006 from the globally respected *Economist* magazine assures us that everything is going along swimmingly.

But there is one, further, pressing issue, summarized in the line, "How ya gonna keep'em down on the farm, now that they've seen Pareee?". In short, for many people through much of history, life on the farm has often been dreadful: low kudos; low income; no prospects; endless drudgery; or as the Chinese put the matter, "Back to the sun, face to the earth". In cruder, economic terms: it is far more expensive to provide people in far-flung places with the basics of life—medicine, education, entertainment—than it is when they are focused in cities. So although the countryside generates less wealth than the cities can (because the modern economy demands that agricultural produce should be as cheap as possible), it is far more expensive to run. Besides, there has been many a survey to show that farmers worldwide are horribly depressed. The suicide rate is enormous (in South Korea, New Zealand, India, Britain). People are leaving the land in droves, especially young people. The average British farmer is now reckoned to be between 54 and 58 years old. In Penang, Malaysia, I have watched the streams of buses in the early morning taking young women from the local villages (*kampongs*) to work in the spanking new, Japanese electronics factory. Anything seems better than the rubber plantation. Surely it is high-handed in the extreme—not simply nostalgic and elitist, but positively Fascist, in the spirit of Franco or Salazar (and some would put de Valera in this category)—to suggest that people *should* stay on the farm, when they so obviously don't want to?

Here is yet another crucial issue that ought to be the subject of urgent global debate, and isn't.

Can the new agrarianism be made to work?

Nobody to my knowledge is discussing this key issue—how to make agrarianism agreeable. Partly, I surmise, this is because it has not occurred to the powers-that-be to address it; and it has not occurred to them because they take it to be self-evident that a city life, with collar, tie, mortgage, Porsche, cappuccinos, and portfolio, is superior; the thing that every right-thinking human being must desire. Partly too it is because the countryside in its industrialized form—all those lovely big machines, industrial chemistry, factory farms, and biotech—is very big business indeed. Farmers farming traditionally, and wanting to be paid, simply clutter up the place. So the matter doesn't occur to the powers-that-be—for the Freudian reason that they would rather not think about it.

Many surveys and statistics do indeed suggest that the world's farmers are an unhappy lot and that their families take every opportunity to leave. But some other surveys, perhaps more sensitively taken, and a great deal of common experience, plus a huge swathe of the world's literature from all centuries and from all countries suggest that rural life can be very agreeable indeed: the most enviable there is. Modern prejudice has it that when town people say they would like to work on the land, this is merely misplaced romanticism, whereas any restlessness perceived among country people is taken as yet more evidence that cities must be superior. I have met farmers in India and Malaysia in tears because they fear they will be forced from their land; and many of the ageing farmers I have met in England cling to their farms even though they are losing money year by year, still trying to farm to the highest standards, just because they love it so much.

In truth, to be happy on a farm you need a list of basics which, in principle, is the same as in any other calling. You need the right temperament—of course farming doesn't suit everybody. You need an interesting set-up. The keenest farmers get bored if in order to make a living they have simply to follow the pesticide manufacturers' instructions and to meet the rigid but whimsical standards of the supermarkets, or cut new grooves in the rubber trees. You need to know what you are doing and what you are trying to achieve. You don't need to be ridiculously rich—the desire for unlimited wealth should be recognized as psychopathology—but you do need a reasonable basic standard, and to feel secure. Perhaps above all, since human beings are such social creatures, you need to feel valued. Nobody likes to be looked down upon.

But everything, it seems, in the modern world, is stacked against the farmer. People born on to a farm commonly cannot escape from it however much they hate it. Contrariwise, in most of the world it is extremely difficult to start your own farm unless you have already made your fortune in some other sphere (and hobby farming is now a significant force in western agriculture). The work on traditional farms can be mercilessly hard—and although mechanization helps in some ways, it makes things worse in others, because it can make the work so tedious. People who spend days on end locked in an air-conditioned tractor cab with headphones to kill the noise, are prime candidates for depression. Traditional gangs at harvest-time worked hard but had a whale of a time. For East-end Londoners hop-picking in Kent was an annual holiday, to be looked forward to: it was the sociality that appealed, as well as the fresh air. I have met modern chicken farmers who have to steel themselves to walk between the rows of screaming birds. If, at the end of this self-immolation you don't earn a living wage, and your children look at you with reproach; and if you are despised because your sun-tan is a little too real and you don't wear a collar and tie; and if indeed you remain uneducated for lack of opportunity and are perceived as a hick; then life is desperate indeed.

Yet it doesn't have to be like this. When farms are properly mixed and geared to the local landscape and climate, they become extremely absorbing: nothing more so. I know farmers obsessed by their olives or their pigs or their vines. In the evenings they just stand and look at them, and dream of even better ways of doing things. If the farms are labor-intensive, as they must be to maintain the highest standards, then each farm becomes a community – and traditional villages, each a focus of several farms, are among the richest communities that humanity has devised. Villages provide the best environments for raising children. In traditional villages, children interact from day one to the full gamut of humanity: old and young, nasty and nice. It's all very educative.

Present-day farm machinery and other high-tech is geared to the big farm—but it doesn't have to be. Appropriate technology is what's needed: appropriate to the small, mixed farm. Some of this will be low-tech. There is a great deal still to be done to improve the efficiency of animal power: kinder harnesses for donkeys; easier axles for carts. There is much more to be achieved with simple systems of irrigation; judicious inter cropping to reduce pest attacks; and so on and so on. But some of the technology most appropriate to the small farm is of the highest kind—indeed it is often the poorest people who can make best use of the highest tech. The same ingenious tools that now make life easier for the suburban gardener or the intensive horticulturist could take a lot of the heart-ache out of small-scale farming—like the cultivator, or computerized irrigation that doles out water drop by drop to the plants that need it most, and solar and wind power. But agriculture is basically a craft industry. The role of science should be to enhance the craft—not, as now, to replace it with rural factories. We might add "Science-assisted Craft" to the desiderata of Enlightened Agriculture and New Agrarianism.

More broadly: there can be no technology more valuable to a new agrarian society than modern hi-fi and the internet. With these, you can live in the remotest corners and yet be in touch with all the world: weed the fields while listening to lectures from Harvard and working for your PhD, if that's what you want to do. As for the services that need to be more hands-on—clinics, local schools: well, society as a whole just has to pay for them. The world *needs* agrarianism, and has to pay what it costs. Simply to decide that proper farming is too dear—or provides too little opportunity for rich people to become even richer—is to sign our own death warrant.

Huge social and economic changes are needed too. Present systems of land inheritance and tenure often militate against all good sense. The "Napoleonic" system, whereby farms are divided among the farmers' sons, and then further subdivided among the grandsons, and so on and so on, are widespread and are an obvious nonsense. The huge land and house prices in countries like Britain, which provide rich urbanites with second homes while dispossessing the people who actually work, are at least equally absurd. When all is done, however, some discrepancy between rural incomes and urban incomes seems inevitable. For example, if 50 percent of people work on the land then the price of food becomes enormous—if the farm workers are earning as much as the city dwellers. So, just as the world needs to work out the just and sensible apportionment of labor between town and country, so it needs to work out a just and sensible apportionment of income. Market forces, which are the modern solution to everything, just don't work: the misery that is already resulting is obvious (except to the people who are doing well out of it, who unfortunately are the ones who make the rules). In general, it would be grand if in all societies people could move freely between town and country so that the people who work on farms or in cities really want to be wherever they actually are. Indeed it should be easy for people to shift between town and country at different times of their lives. It is exciting to be in the country when you are a child—children brought up on farms are generally to be envied—and becomes highly desirable again when you become a parent. But if people want to take themselves off to the city and generally to see the world between the ages of, say, 16 and 30—well; that seems eminently sensible. But then, if they choose, they should be able to come back again. At present in most circumstances this is all-but impossible.

So of course the new agrarianism raises enormous problems. But all, in principle, are soluble. The first requirement is to address the problems, because the new agrarianism is necessary. No other route makes sense. The fact that the powers-that-be are not addressing these issues, but have merely concluded that farming must be an anachronism because so many farmers are poor and seem unhappy, is yet another indictment of their poverty of thought; yet another reason why humanity as a whole has not only to think seriously about what it needs to do to make life agreeable and indeed to stay alive, but also how to re-organise its own governance. We cannot leave our future to people who just aren't aware of what is important, and have no intention of changing their minds.

So what are we going to do? I discuss at least the beginnings of change in the next chapter.

Chapter 6

RENAISSANCE: THE WORLDWIDE FOOD CLUB AND THE COLLEGE FOR ENLIGHTENED AGRICULTURE

Wwhat can the rest of us do when our leaders, the people who most influence events, have lost the plot?

The world these days is driven by a great gyre—a positive feedback loop of money and technology. Big companies—corporates—produce wealth. That is what they are for. The wealth in turn finances science. The science in turn produces "high" technologies—technologies of the kind that depend on science. The new technologies provide new goods for the companies to sell—or in the case of iPods and so on the new technologies *are* the goods. So the companies become richer, and so they can spend more on science, which then can generate more high tech, and so on and do on.

The loop runs itself. It is self-reinforcing. Modern western governments typically use tax-payers' money to finance the basic science that feeds the really big ideas into the loop. But then they are content to stand back and watch, and feed upon the wealth that gushes forth.

The wealth that does gush forth has produced some spectacular results. Many modern cities are indeed fabulous. Modern supermarkets at least in favored places are cornucopia, or so they seem. They would have impressed Cleopatra, though they are available to everyone. Modern intensive care units really do save lives. And so on.

Yet there are obvious snags. The commercial companies on whom the whole system depends are obliged in this competitive world to be as profitable as possible. If they are not, their shareholders will invest elsewhere. The companies can make pleasant gestures for the sake of public relations—hand out computers and footballs to primary schools—but they cannot afford to be altruists. More broadly, they cannot afford to operate as public services, much as modern governments might like them to. Overall, since they are obliged to make money, they must cater for the rich, who have money to spend, more than for the poor.

In practice, then, this fabulous gyre, which generates such wealth and spectacle, spends much of its energies on technologies and enterprises that don't really matter, or indeed are destructive (such as those SUVs, sports utility vehicles, that when fully revved do one mile to the gallon and now account for nearly a quarter of US car sales). But it often spends very little on things that really do matter. Within agriculture, billions are spent on pesticide research. Very little indeed is spent on the methods of biological control that would make those pesticides unnecessary. Yet the world's position is now so bad that people who are normally reckoned to be conservative—senior scientists and archbishops—now tell us that we have at best an even chance of getting through the present century in a tolerable condition.

Worst of all, though, is that nobody seems to be in control—not even the governments and corporates. A senior executive in one of Britain's leading supermarket chains told me recently that her company would like to behave better than it does but could not—because better behavior would be expensive and its chief rival would cash in. That rival would doubtless say the same. Modern politicians of the Tony Blair stamp say they cannot interfere because the gyre creates wealth and they cannot afford to kill the goose that lays the golden eggs. Forty years ago President Nixon spoke of the commercial-technical engine as an "animal". A wild animal, on the rampage.

In short, the governments, corporates, and their expert advisers really do have huge influence and yet they are not truly in control. When challenged, they are wont to tell us that they cannot really be held responsible. The world's economy is like a great ship, ploughing across a vast ocean with tremendous drama and a prodigious wake but with no particular destination in mind. The powers-that-be are stoking the engines. But in the end they just are passengers like the rest of us. Or so they tell us, when challenged.

So we need a new system—and we need new governance. Since the powers-that-be will not or cannot deliver what the world really needs, we, humanity, must dispense with them and assume the governance ourselves. These comments apply generally. But here I want to focus on the world's food supply chain. This is surely the best possible place to start because the food supply chain is the thing we really have to get right—and if we do get it right then a great deal else that matters can begin to fall into place as well.

So how?

How can we take control of our own affairs?

Overall, in principle, there are three possible ways to change the status quo:

The first route to change is by Reform: persuading the powers-that-be (whoever they are in whatever context) to change their ways. This, I suggest, holds very limited promise. Reform brings change incrementally; and it is very hard to see how a multi-national supermarket, designed expressly to minimize costs (by buying as cheaply as possible); to "add value" (maximum meat; out of season fruit from the far corners of the Earth); to maximize its own income (by persuading people that its extremely expensive produce is in fact cheap); and to direct the resulting profit to the company's directors and shareholders (as opposed to the producers), can be transformed incrementally into the kind of institution that truly could foster Enlightened Agriculture and the New Agrarianism. The American biologist Sewall Wright made the same point in the context of evolution. In principle, he argued, it would be possible to turn an elephant into a daffodil, just by changing it gene by gene. In reality, however, a creature that was half elephant, half daffodil, would be non-viable. So the step-by-step, incremental transition could not in practice be made.[1]

---

[1] *Sewall Wright was a good Darwinian, and he was not arguing against the possibility of gradual evolutionary change in general. Neither did he want to argue, as some modern Creationists do, that complex organs could not arise step-by-step. He was merely pointing out that in practice, some transitions are not possible.*

The corporates that run modern supermarkets are global traders who also contrive to control both the food processors and the farmers (it's called "vertical integration"). What the world really needs is what it traditionally had—Adam Smith-style markets that bring together many different traders, mostly local and all doing their own thing, either in the open air or in some market hall of the kind that can still be found in many a town center. In short, a world food supply chain based on Enlightened Agriculture, operating within a culture of New Agrarianism, does not need globalized corporate supermarkets. Corporate supermarkets and all that goes with them are indeed the opposite of what it needs. It merely needs market halls where the mostly local producers (and bakers and pie-makers and so on) can sell their wares. The only thing the corporate supermarkets have to offer a world that really was geared to feeding people well, is its real estate.

The corporate superstructure, all those directors and young execs with their MBAs, are entirely superfluous; as, too, are all those very clever but misguided young men and women devising ever more ingenious TV dinners. A safe and agreeable world is one in which people at large can cook, and people who can cook don't need their meals ready-made. To revert to the evolutionary analogy, the modern supermarket is a metaphorical elephant while what the world needs is a host of metaphorical daffodils. There is no plausible route from one to the other.

By the same token, governments that depend on corporates for finance, and which above all seek power, have nothing to offer a world that has moved away from the whole idea of hierarchical government. So governments of the modern kind cannot usefully be reformed either. I know plenty of people—very good and clever people—who spend much of their lives trying to persuade corporates and governments to change their ways: to do things ever so slightly differently. I don't want to say that they are wasting their lives but deep down, that's what I feel. An elephant can become a slightly different elephant but it cannot become a daffodil. Sooner or later in the chain of incremental change steps need to be taken that are implausible.

Reform does have some role to play, however. Consumers can mend their ways. People who can't cook can learn to cook. Children brought up on turkey dinosaurs and frozen pizzas can learn very quickly that this is not how food needs to be. If people at large re-learned how to cook, and to appreciate food as the peasants of southern Europe and India and Turkey have always done, then they would see once and for all, without being told, what nonsense we now put up with.

The second well-tried route to change is through Revolution. For various reasons, I do not recommend this. For one thing, on the large scale at least, revolution is not going to happen. Those who want to bring it about will again waste their lives and in these days of trigger-happy governments, who apparently seek to rule by martial law, anyone who protests too obviously is liable to be shot. There seems no point in this at all.

For another thing—more profoundly—revolution on the scale undertaken in Russia in 1917 and in China in the late 1940s is not called for. The world does not need centralized economies of the kind that the organized Communists set up. Capitalism can do the trick. We simply need to rediscover the moral as well as the practical roots of it—as defined very brilliantly by the founders of the United States, not least by Jefferson and Madison; and to explore modern, benign models of it, as outlined in the last chapter. As I keep saying (but it can hardly be said too often): it isn't capitalism *per se* that is killing us off. It is the modern, hyper-extended, simplified and brutalized version of it. It is of course hugely unfortunate—possibly fatally so—that the brutalized version of capitalism now dominates, for the crude logistic reason that it is designed expressly to dominate. It does not waste time on the niceties that exercised the men and women of genius who created the modern United States. We need to fight the modern perversion of capitalism, to be sure, but a Maoist head-to-head conflict is not the most appropriate way to go about it.

Finally, revolution is extremely risky precisely because it is revolution. Once begun, it is very hard to control. In truth, we cannot tell where any concerted human action will lead: chaos theory always applies. But the outcome of revolution is even more unpredictable than with most human endeavours. In reality, the chaos engendered tends to be controlled eventually if at all by despots—which is not, generally, what the revolutionaries themselves first envisaged. So let us not waste time on the revolutionary route.

There is a third possibility. Renaissance.

Renaissance of course literally means re-birth: starting again. In the present context—we, humanity, need and want a new food-supply chain. So let's just create it. We don't need to ask the governments for permission. We don't need to persuade the corporates to do it for us. We just need to do it.

In truth, the process of Renaissance has already begun. Many of the groups who are seen as reformers, and present themselves as reformers, are not, in fact, simply trying to change the status quo. They are already creating new ways of doing things—or making sure that traditional, necessary ways, continue. Indeed, as I stress in all that follows, the Renaissance that the world requires can at least be put in place simply by bringing together the many different initiatives that are already in train. I do not claim to be particularly original. Other people have already put most of the necessary components in place, and done most of the necessary thinking. There is much more to be done, to be sure, but the prime task is one of coordination. Neither is it necessary, in order to bring about crucial and permanent change, to persuade the majority of what needs doing. All that's needed is a critical mass—and a critical mass can be quite small, at least to begin with. Majority (and hence, true democracy) can come later.

In this endeavour, too, there is yet another wondrous serendipity. The thing that we most need to get right—the food supply chain—is, in reality, the thing that each of us can most directly influence. Not everyone can be farmers, but we all can support those who are—I mean the ones who do it properly. Not everyone has access to land where they can grow their own food, but many people have. And everyone can cook. If you don't have a kitchen then, *faute de mieux*, get yourself a butane stove.

Because in principle everyone can cook, everyone can be a gourmet. There is nothing like cooking to encourage appreciation. French, Italian, Turkish, Lebanese, Indian, Chinese, and indeed English and Scottish and Welsh and New England and Mexican kitchens are still stuffed with gourmets as accomplished and refined as the greatest of chefs—as the greatest of chefs acknowledge. In every field, the people who are truly great never stop learning.

Cooking is the ultimately anarchic act. You do need good ingredients but in general (give or take the odd bottle of soy sauce or Tabasco or whatever) the simpler the ingredients, the better. If we could all cook—or if a critical mass of us rediscovered the joys of it—then, I reckon, the whole sorry superstructure of the present corporate-government-bureaucrat-technologist food supply chain would begin to fall apart. It relies upon ignorance, and the general sense of disempowerment. Cooking is not generally taught in schools these days, and apartments are built without kitchens and in many a major city you can walk from one end to the other and never catch sight of a good loaf of bread or a fresh vegetable—nothing but pizzas and burgers and fried chicken—and modern governments don't seem to think this is anything to do with them because although they interfere in our lives at every turn they do not seem to think that their job description requires them actually to *govern*, and to make the world agreeable. Above all, modern governments will not take on the corporates who run the world, including its food supply. On the contrary: they keep their jobs by supporting the corporates, which are the fountainhead of wealth and hence of power.

In short, a prime task is to establish, worldwide, in every country, a food culture: a critical mass of people who really appreciate food, and will put themselves out for it. This is not just an exercise in self-indulgence—far from it. As I pointed out in chapter 3, great cooking, sound nutrition, and the best kind of farming that is founded in kindness and good biology go hand in hand. They are an inseparable trinity. A universal food culture is necessary, the *sine qua non*: and it is the thing that *everyone*—even the most apparently disempowered—can help to bring about. Such a culture may well in practice begin (or begin again) with the middle classes, and critics will accordingly suggest that it is elitist. But as Lenin said, all revolutions begin with the middle classes (and we are not at this point speaking anything quite so radical as revolution). Let us begin where we may. But get it right. Out-of-season strawberries and pallid and over-puffed chicken breasts just will not do.

Yet the food culture, though vital, is not enough by itself. My grandest practical proposal for the world's renaissance is what I am calling the Worldwide Food Club.

The Worldwide Food Club

The Worldwide Food Club is conceived as a cooperative of people at large who really care about food—truly informed consumers—and of food providers who truly desire to supply it: producers—farmers and growers; and preparers—cooks, brewers, bakers, butchers, charcutiers, picklers, caterers, restaurateurs. The emphasis is on "cooperative". The club is not conceived simply as a consumer movement, putting pressure on providers to bring their prices down. It is not a cartel of providers, controlling what can be bought and gulling people into paying more, which is the modern supermarket way. It is a pact. The providers undertake to do what they do to the highest possible standard (which is what, in my experience, many truly desire to do); and the consumers who give a damn, undertake to supply them with a steady market, through thick and thin (which is what all providers in all fields need). The emphasis, too, is on "club". The members—consumers, cooks, bakers, farmers and so on—decide who should belong, and what the standards must be. But it is not an exclusive club. Anyone can join. The only requirement is a dedication to excellence—not simply in the food itself, but in the underlying morality: it matters how the food has been grown, and where, and by whom, and who profits by it. Information on the Club can be found at http://www. parinetwork.org/site/?cat=10

Organizationally, there are many precedents for such a club. Indeed, all existing clubs of all kinds provide the precedent. In clubs, the members decide what happens, and who belongs. If there is a ruling committee (as there usually has to be) then that committee is as Abraham Lincoln envisaged all government should be—of the members and for the members; and if committee members fall short, or start to line their own pockets, then in well-regulated clubs they can be deposed immediately. Only in nation states, such as Britain, do the people have to wait for half a decade before they replace somebody they dislike with somebody they dislike slightly less. Clubs meet the criterion of democracy that Sir Karl Popper defined: what counts is not how or even whether the people elect their leader, but whether or not they can get rid of them. By that criterion, the most powerful alleged democracies in the present world don't even register. Trusts, too, provide a precedent. Trusts in general generate wealth and have a central pool of wealth—but the wealth belongs to no individual: only to the trust collectively. It is overseen by trustees, who are unpaid. They are merely trustworthy. Again we see that existing structures—in the case of the trust, essentially capitalist structures—can serve the world's purposes. A future world that works has to be very different from the present one but that world can to a significant extent be created from elements that exist already. Renaissance involves serious rearrangement, but does not require that everything be reinvented.

Quite soon, though, such a cooperative of like-minded suppliers and consumers, content to deal with each other, could become a significant force in the world as a whole. Eventually—indeed quite quickly—it could begin to supplant the present powers-that-be. Present-day supermarkets, with their surface pzazz and their underlying destructiveness, would fade away simply because people at large could see better ways of doing things. There is no need for an out-and-out fight. They can be left, as Lenin put the matter, to wither on the vine; and so too, more broadly, would the present style of governments who depend absolutely on corporate wealth, and adjust their policies and manipulate our lives accordingly.

Finally (before I move on to the practicalities), although I have suggested that cooking is "anarchic", I do not envisage that a world that works around and otherwise ignores the present powers-that-be is innately anarchic. Coordination is necessary, and coordination requires some kind of organization. Supporters of the free market argue that the market itself provides all the coordination necessary—it should, after all, ideally, give rise to an "invisible hand" that will in the end produce a benign result—but it is already obvious that it does not and cannot deliver the social benefits that are claimed for it. It just won't do. But organization need not imply the kind of top-down hierarchies that we have now, dominated by people dedicated to power and to the creation of wealth that reinforces power, and by intellectuals and experts who are content to go where the money is. There are other models. One, in general terms, is the "neural net": conglomerations of equal players who self-organise for *ad hoc* purposes, in ways that are good for individuals but never, like the modern-day corporates, acquire enough power to unbalance and redirect the whole world. I am aware that neural nets are complicated things but in principle they are what we seem to need, and the complications just have to be worked at. The New Agrarianism is complicated too, but it is necessary, and the task is to make it work. The spirit of the Worldwide Food Club—competent people relying on their own craft and cooperating to create a benign society, with no individual striving to be outlandishly rich or to dominate the rest—seems to me very much in the spirit of Mahatma Gandhi. The global free market will not do. But the far more subtle and deeply rooted philosophy of Gandhi will do very well.

In practice, I see the Worldwide Food Club evolving in four phases. Thus:

Phase 1: The Global Exchange

The initial stage of the WWFC can be found at the Pari Network website. In the next stage the website will list people who want to be involved: an ever-expanding, ever-evolving directory: an attempt to put like-minded people in touch with each other. These lists should not be simply of names and addresses. Each listing should be a mini-portrait (as far as possible, with illustration); and of course should cross-refer to the person's/institutions' own websites.

Emphatically not is it the intention of WWFC to reinvent the wheel. As I have already intimated, there are endless excellent initiatives already in train. But they are of many different kinds and may seem, at first sight, to be pulling in different directions: some practical, some political, some informative, some campaigning, some tiny, some global, and so on. The idea is that between them these many and various movements could create a critical mass that truly could set the world on a new course, if only they worked together more than they do. "Critical mass" does not mean "majority". Critical masses in the short term can be quite small. But once they exist, they begin to make a serious difference.

The listings in the directory fall into several categories, as follows:

List A: The sharp end

1: Producers—farmers, market gardeners—who are producing food of the very best standards by the most acceptable means, or who would like the opportunity to do so: farmers who are locked into modern commercial systems that they know are foul, and seek to escape; ex-farmers, who have been shoved out by modern politics, but want to get back in; or indeed townspeople, who yearn to be farmers.

I vouchsafe that there are millions—actually, hundreds of millions—of producers and preparers out there who long to do things differently, and better, than they are being forced to do things at present. I have not spoken to all of them, but I have met many hundreds of farmers, cooks, and so on, in all the habitable continents and on sizes of farm from the minute to the vast, who are thinking exactly along these lines.

2: "Preparers" (there seems to be no collective term except for "processors", which has the wrong connotations): millers, bakers, charcutiers, butchers, brewers, wine-makers, distillers, cooks and caterers of all kind, who want only the opportunity to work to the highest standards.

3: Consumers. The consumers are the drivers. The millions of farmers and preparers who yearn for excellence often cannot do the things they know should be done because they have no guaranteed market. Increasingly they are forced to sell to supermarkets, or to various processors or middle-men who supply the supermarkets, and these manufacturers and traders are interested primarily in keeping prices down. Their standards—apart from price—are entirely to do with the criteria of marketing and distribution: size and shape, color, shelf-life, fashion. Yet, as is abundantly clear, there are many millions of people out there who want to buy good things. They are the potential market. All they need is opportunity: ease of access to what is good.

In the first instance, the WWFC website will list all the producers/preparers who are already producing good things; with an invitation to all who log into the website, to get in touch with them. At its simplest, then, the WWFC website is a shop window for farmers, brewers, bakers, right-thinking restaurants, and so on.

List B: Box schemes and other existing trading exchanges

In addition to the producers/preparers themselves there are many existing schemes designed to channel produce from several or many producers to people at large. These include companies such as Britain's Real Meat Company; box schemes of all kinds for delivering local produce; and so on.

List C: Collaborators/sympathizers

An annotated (and illustrated) list of all institutions, both tiny and huge, those ambitions are broadly similar to, and are known to be sympathetic to, the general cause of the WWFC. In Britain, these would include: The Soil Association; The Slow Food Movement; The Fair Trade Organization; The Green Party; The Food Animal Initiative; Compassion in World Farming; The Cooperative Society; The Quakers (and, potentially, many other religious movements); The Campaign for Real Ale; The Food Ethics Council; and many more. But the C list, like all the lists, should be international.

List D: Information

The WWFC website should also be an exchange of pertinent information. This can take two main forms:

I: There could and should be an element of newsletter, alerting loggers-in to new events, newcomers to the list, etc.

II: References to other information, including useful books and websites.

Just to re-emphasise: Some people I have already spoken to have suggested that various of the above organizations already do all or most of what I am intending to do with the WWFC. On the one hand this is abundantly true—for if it were not so, the WWFC could not get off the ground. Yet it is *not* the case that any existing organization achieves, or is intended to achieve, what the WWFC is trying to do. Almost all of the existing organizations have an *ad hoc* function: for example to promote organic farming specifically (which is the core thrust of the Soil Association); or to campaign for animal welfare (as in Compassion in World Farming); or for more equitable trade (as in the Fair Trade movement); or in a more general way to talk about ethics (as in the Food Ethics Council).

Some existing enterprises are cooperatives of producers, or are consumer movements of various kinds. Yet very few indeed have the agenda of the WWFC—which is to create a truly collaborative and cooperative movement of *all* players in the global food supply chain: all, that is, who are driven by concern for all humanity, and other species, now and in the long term. The Slow Food Movement may look and sound similar and it should indeed be an invaluable ally but its role is specifically to create and reinforce food cultures. This is essential, for the world cannot create a viable food supply chain unless it is pulled along by real enthusiasm for food and a knowledge of it. But the grand task is to match the food culture with suppliers able to meets its needs, and to ensure that the transactions are fair to all concerned, and carried out with proper concern for the landscape at large and for the future. Perhaps of all existing institutions, the Co-op comes closest to the WWFC. Even so, the Co-op exists primarily as a commercial trading company (although it also has its own political party). It is not expressly intended or designed to create an alternative food supply chain to serve all humanity.

The WWFC in contrast to all that exists so far is conceived as a network—a meta-, or a virtual institution, intended to coordinate and thereby enhance the efforts of everyone along the food supply chain, and all who take a serious political interest in it, who are already doing the right things.

A great deal could emerge from Phase 1: new trading patterns and coalitions that could be of great though as yet unforeseeable value to a great many people. But sooner or later (and probably sooner) it will be useful or indeed necessary to introduce some structure into the system.

Phase 2: Infrastructure

There seem to be two obvious requirements.

First, the website itself will need to be maintained—which would take some hours every week. Volunteers might do much of what is required, but some kind of paid staff seems necessary too.

Yet it is essential that the WWFC should maintain its status as a non-profit Trust: a service to humanity, though of course paid for by humanity. So it will need a board of trustees: not a ruling body, but a group to ensure that principles of justice and good sense prevail.

Secondly, Craig Sams of the Soil Association has pointed out that farmers are not too keen on publicizing their own efforts and marketing their produce. They tend not to see it as part of their job. In truth, part of the point of the WWFC is that farmers should not have to turn themselves into butchers or grocers in order to make a living—there is no more important job than farming and farmers should be properly paid for being farmers. Neither should the WWFC as currently conceived do their marketing for them. Nevertheless, many farmers do need more promotion; and to this end at least, the WWFC should be able to offer advice and practical assistance. In short, WWFC cannot long remain simply as a shop-window and clearing-house. It needs brains and know-how behind it to make sure that the shop-window is well stocked and that all sides are well served.

Website staff and competent advisers need funding. So this too will soon become a requirement. There are many possibilities, including grants, donations, and subscriptions. But these practicalities are up for discussion.

Once fund-raising is in train the WWFC can begin to move into Phase 3:

Phase 3: Material Presence

Truly to fulfill its mission—to provide an alternative food supply chain for the whole world—it would be at least highly desirable for the WWFC to acquire a physical presence, to support its meta- or virtual role. Sooner rather than later, the WWFC should acquire land of its own where good farming practice is ensured; and permanent marketplaces of all kinds where everyone can buy good food in the traditional, over-the-counter ways.

All WWFC property should be held in trust, as with the Sierra Club or the Audubon Society. The farms would either be run by tenants—free to do their own thing but operating nonetheless within the WWFC rules and guidelines; or they could be run by managers. Many farm managers have a great sense of mission and are excellent, for people on salaries can work just as conscientiously as owner-occupiers if they believe in what they are doing. The farmers (or managers) would operate as commercial units. But at least the core of their market would be guaranteed by the WWFC itself; and although the WWFC's tenants would pay rent, this as far as possible would be peppercorn—nothing like the mortgages that oblige so many existing farmers to farm purely for cash, whether they like it or not.

Of course, many excellent farmers exist already—for if it were not so, WWFC could not even begin. More urgent, then, is to acquire marketplaces. In physical form these could be supermarkets, or traditional market halls, or indeed small village or high-street shops, possibly run under franchise, or simply in partnerships. Whatever form they took, they would function as permanent farmers' markets, focusing on fresh local produce and/or on produce shipped in from further afield but always under trade agreements that are good both for the producers and for the environment.

One practical advantage of existing supermarkets is that although customers can load their baskets from many different shelves and stalls, they queue up only once to pay. In a traditional market the customers deal directly with many different traders (with all the Adam Smith-like and social advantages that this entails) but also have to queue and pay several times over. With existing technology it would be easy to combine the advantages of both; shop where you like from whom you like but then pay only once at the end. When all the customers have gone home the barcodes can work out how much of the proceeds belongs to each trader.

By all these means, possibly in very little time—less than a decade, perhaps—the WWFC could become a serious presence in the world: the inspiration for good food and good practice; and the necessary conduit between producers and customers. The WWFC must pay its way, as all institutions must, but it is not designed to maximize profits or to make rich people richer. No individual member of WWFC owns any part of it. It is a collective, run for the benefit of humanity as a whole. There have been many such institutions in the past. They depend upon basic, shared, human decency and trust—but of these there is no shortage.

This leads us into Phase 4:

Phase 4: A New-created World: Democracy at last

The WWFC, when it begins in earnest, will be true democracy in action. Its policies and its *modus operandi* will reflect the will of all its members. Democracy requires that human beings should be cooperative and unselfish, with a true sense of community; that they are prepared to trust each other;

that they do not regard truth as a bargaining device and betrayal as a tactic, or feel the need to protect themselves obsessively against the treachery of others. Democracy depends, in short, on an optimistic view of human nature. But in this, I believe, we have very good reason to be optimistic.

One more new kind of institution is needed: a College for Enlightened Agriculture.

The College for Enlightened Agriculture

The task is not simply to create a new, democratic food supply chain, vital and momentous though this is. We also need to ensure that we do not allow the world to get into the same kind of mess again. To do this, we need to ensure that the food chain and the world in general cannot again be taken over by particular vested interests and power-groups, whether political parties or vast commercial companies. There can be no guarantees of permanent security, but we can seek to ensure that the new food supply chain is as robust as possible. This requires that we should as far as possible anticipate and think through all the difficulties that might arise. This requires continuing, hard thinking by people from a whole variety of different fields, from farming to soil chemistry to economics to moral philosophy—everything is relevant. By definition, people from different fields acting together to a common theme form a college. What the world needs, I suggest, in addition to the Worldwide Food Club and providing its intellectual base, is the College for Enlightened Agriculture.

The kinds of questions that the members of the College should address must unfold as the thinking proceeds. But there seem to be two essential classes of questions: those that have to do directly with food and agriculture, to help us work out the practicalities of Enlightened Agriculture; and those that are concerned with infrastructure—the kinds of economics, politics, science, and moral and aesthetic attitudes that are needed to create the kind of world in which good farming (the kind that it intended to feed people) can be practiced. So here are two preliminary shortlists:

I: Questions of Food and Agriculture

1: A worldwide survey of traditional farming and cooking

There is no doubt, at least among people who have actually looked, that the wisdom of farmers and cooks that has accumulated this past 10,000 years and more is prodigious, and often brilliant. Modern agricultural science has apparently succeeded as well as it has only because it had such a broad and deep base of craft and traditional knowledge to build upon. But that craft and knowledge is being swept aside as fast as the powers-that-be can arrange. The general attitude was summarized for me a few years ago by a young MBA who assured me that "the world began in 1980".

We need to rescue what is left of what people know, and their skills, and their reasons for doing the things they do, before it is all gone. Enlightened Agriculture is not an exercise in nostalgia—emphatically not. But to a significant extent the adage applies: "The future lies in the past!"

2: Can organic farming feed the world?

This is a key question: the thorough exploration of it must hugely enhance our grasp of what needs doing, and what is possible. Enlightened Agriculture is not simply an exercise in organic farming. But organic farming at least represents the baseline, the default position; and it is very important to know whether and to what extent the superstructure of industrial chemistry and biotech—the superstructure that now drives the whole system!—is actually necessary. The details are vital. To what extent can we control pests and diseases by biological means—the right crops in the right places, rotations, mixed cropping and intercropping, encouragement of predators, and so on? Can we maintain the necessary yields just by nitrogen fixation and recycling? The theoretical answer must be "yes"— all the nitrogen and phosphorus that we remove from the ground in the form of food reappears later. The only question is whether it is practical to get it all back on to the land—which of course raises many more issues of a practical kind.

People are working on all these questions, to be sure. But the powers-that-be who have most of the wealth and wherewithal are much more interested in other things—how much fertilizer and pesticide is it possible to persuade farmers to buy? Or, these days, how much profit is there in persuading farmers to grow GM crops which in theory (though not necessarily in practice) require less chemical input? Which course is the more profitable? But these are not the questions humanity needs to be asking if we are serious about our future—if, that is, we give a damn about our children.

3: To what extent can all the various countries in the world truly become self-reliant in food?

As I have stressed, "self-reliant" does not mean "self-sufficient". Self-reliant means producing all the food at home that the people need to get by on so that they have autonomy. It does not mean producing everything that is desirable. Britain, for example, should not be trying to produce bananas, except occasionally for fun in botanical gardens (and so far as I know Britain never has tried to grow all its own bananas). This means there is plenty of scope for trade within enlightened agriculture— provided only that a few obvious preconditions are met. That is: transport of exported and imported goods should not use disproportionate amounts of energy; the trade must be fair, so that the producer countries, and in particular the producers themselves, do well out of it; the commodity crops for export must be grown without wrecking the ecology of the producer country; and those crops should not in general be grown at the expense of self-reliance (as has almost become the norm).

But at this moment, no one can say whether and to what extent any one country *could* be self-reliant in food. Some countries that have experienced famine in recent decades could surely feed themselves several times over if only the indigenous agriculture was given a lift—if indeed, it was not systematically undermined, and besieged by war. Angola comes readily to mind—a country with brilliant traditional farmers and every kind of climate which, nonetheless, has been abject for much of the past 30 years. What is true of Angola is probably true of all other countries in Africa as well. Ethiopia has at times been a bread-basket. Other countries—Japan is an obvious example—would probably be hard-pressed to achieve self-reliance. But we need figures, data, projections that take global warming into account. It is absurd—negligent in the extreme—that such studies are not done, seriously and continuously. Instead, agricultures worldwide are systematically wrecked in the name of progress that is defined in terms of GDP and visible wealth. This just is not good enough.

4: Biofuel

No one doubts that biofuel must play a greater role in the world's affairs as fossil fuels run out. It would be a tragedy, however, if farming for food found itself in head-on conflict with farming for diesel. Judging by the events of the past 30 years, and the present gung-ho pronouncements of some oil companies, it is a tragedy we seem most unlikely to avert. If agriculture and the world in general continue to be driven by the perceived need to maximize profit by whatever means, and to maximize GDP, then it is obvious which way the scales will fall. At present, in straight contests between soya for European cattle and sorghum and maize for local consumption, the cattle win. In a future contest between local food and the demands of the SUVs, the SUVs will surely win.

Enlightened Agriculture needs, probably, to integrate specific fuel crops. But it needs to stay enlightened. The practical details, as well as the underlying economics and morality, need much more thought.

5: Agroforestry

One of the few truly encouraging trends of the past few years has been the increased interest in agroforestry—combining production of food (or, conceivably, of biofuels and other commodities) with forestry. The variations are endless. Livestock can fare particularly well under trees. Pigs, poultry, and even cattle are basically forest animals. They are demonstrably happier and more productive with shade and shelter. Pigs were traditionally raised in forests—as they still are in southern Spain and Portugal, among the cork oaks and chestnuts. In the tropics, yields of dairy cattle may increase by 30 percent when they have shade. The traditional herds of the American prairie must have been seriously stressed. In Indonesia, Professor Bob Orskov has shown how to raise two head of cattle per hectare on the spaces between oil palms—which is a high stocking rate even by European standards, and benefits the trees as a bonus. Near my home in Oxfordshire, chickens are being raised under newly-planted birches and willows; free-range chickens are far more inclined to go outside if they feel protected from aerial attack. Many crops flourish under trees, too. Tea and coffee do far better in shade. Cardamoms demand shade. In the Mediterranean I have seen broad beans (a fine staple crop) and vines among the olive and fruit trees—it is traditional practice. Agroforesters in the short term can make a living from their animals and crops—and look forward to a timber crop of walnuts, oaks, teak, or whatever; all of which, while growing, provide shelter and food for wildlife and reduce the world's carbon burden. Willows can be continuously cropped as a source of biofuel. In short, the possibilities are immense—and the husbandry is immensely intricate, and therefore immensely absorbing: the absolute antithesis of the farming-by-numbers industrial monoculture that is now called "conventional", and is so unattractive. Good work is being done on agroforestry—but there can never be enough.

6: Farming and wildlife

Farming now occupies about a third of all land, including most of what is most fertile—the lush valleys and coastal strips. National parks, largely dedicated to wildlife, are largely confined to the places that are hard to cultivate. But most wild creatures prefer the easier places, just as people do. Britain's wild cats, golden eagles, national symbols of untamed nature, are confined to the woolliest of Scotland's Highlands only because they have been booted out of everywhere else. Worldwide, farming policies can be just as important for wildlife, as the specific conservation measures. Europe's Common Agricultural Policy is now geared in part towards wildlife but this clashes with its overall ambition to turn Europe into a semi-united political unit, and with its perceived need to compete in the global market. So in this as in most things the CAP is a mess. Enlightened Agriculture includes the idea that farming must be as friendly as possible to wildlife. But much serious thinking has yet to be done to work out the details.

Many more questions of this kind, some very broad, some detailed, will present themselves. Clearly, the College must be very aware of all the people in the world who are working on such issues to avoid duplication, and to effect collaboration—a much more fruitful approach than the competition that has become so fashionable.

But in addition, there are background issues to be worked upon:

II: Economics, politics, attitudes

1: An economy fit for Enlightened Agriculture

Farmers are constantly told they must gear what they do to economic realities. As a generalization this is obvious. But we should ask as a matter of urgency whether the world's prevailing economies are what the world really needs. Since we are heading for, or are already embroiled in, all kinds of disasters, all of which have largely economic roots, the answer obviously is No. So we need new economic models for a whole raft of reasons.

Important among those reasons is that the new economic models must encourage enlightened agriculture, because the world needs enlightened agriculture in the interests of society and of biological reality.

It follows that the new economic models (which the world needs for many reasons) should to a significant extent be framed specifically to accommodate enlightened agriculture (which the world needs if it is to survive at all).

In short: to a significant extent, new economic models and enlightened agriculture should be developed in concert.

(And in general, the way things have been done up to now—devise an economy and a political system and then seek to ram agriculture into it—is absurd, and of course extremely dangerous. These in large part are the chickens that are now coming home to roost.)

2: Can corporates really help to make a better world?

I know many good people who work for corporates, some of them in senior positions. I have addressed the World Economic Forum in Davos, Switzerland, where CEOs meet with presidents to discuss how they can make the world a more agreeable and safer place, and some of them at least are truly intent on finding answers. Corporates make many altruistic gestures and since they are so rich, they can dispense largesse out of petty cash that to most individuals, or even to most NGOs and even to entire societies, seems beyond the dreams of avarice. In Britain, Tesco is currently giving cart loads of soccer balls to schools which, for some reason that in this boom economy of ours must remain mysterious, apparently lack soccer balls. In recent years the company we all love to hate, McDonalds, has promoted some significant improvements in animal welfare. Let us give credit where it's due. How easy it would be, then, and how tempting, simply to invite the corporates, rich and vigorous as they are, to help us solve the world's problems.

Yet there are deep questions to be addressed. How far can a corporate afford to be altruistic before it begins to erode its own *raison d'être*? To take one specific example: to what extent can a supermarket chain support local farmers without undermining its own role, position, and source of wealth, as an international trader, playing one producer off against another, all around the world? Insofar as corporates do good, how much of what they do results directly from their corporate status? Couldn't the same outcomes be achieved by different kinds of institutions, without the drawbacks that corporates seem to bring? Isn't it simply the case that the corporates are now the fountainheads of all benison simply because they have systematically wiped out all the alternatives? Mightn't the alternatives be better, if they were given a chance?

More broadly: corporates are engines for making money. This they do very effectively. Is it really plausible that a collection of such engines, to a significant extent locked in perpetual battle but also with an inevitable element of cartel, can produce a world that is good for most people to live in, and can flourish indefinitely? Can this ritual battle of corporates really produce the best of possible worlds? On the face of things this seems most unlikely. Indeed such an idea seems ludicrous. But before we write off the corporates altogether and make them the enemy, we should give serious thought, and give them the opportunity to state their case (without the rhetoric of PR and general spin). If they *could* fit into the kind of economy the world needs, and indeed help to make it work, then that could make life a lot easier. On balance, at present, this does not seem likely. But the issue is too important to pre-judge.

3: How many people should work on the land?

Ninety percent of people on the land, as in Rwanda, is obviously too many. To anyone with any sense and knowledge it is at least equally obvious that one percent, as in Britain and the US, is far too few. Yet to the powers-that-be, the US is the model in all things.

People who do have sense and knowledge and give a damn should be asking, what is the right proportion of people on the land? Just to set the ball rolling, I have suggested in this book that the minimum in countries like the US and Britain should be around 20 percent (meaning that we are way off beam), while the maximum, in countries that are not yet industrialized (and perhaps in reality never can be), might be around 50 percent. But this needs thinking about. My suggestion is a preliminary guess. But it is a great improvement on what the powers-that-be have so far come up with.

Adam Smith, incidentally, expressly addressed this issue: what is the best ratio of rural people to urban? It isn't a new question. It has merely been forgotten about, like a great deal else.

4: How can we make agrarianism work?

There is a long short-list of problems: how to get medical care and good education to people who are widely scattered; the general problem that rural communities generate least disposable wealth, head for head, and yet can be the most expensive to maintain. And what should be the relative income of rural and urban people—for if there is simple equality, while a large proportion of the people work on the land, this raises all kinds of problems. How can we arrange for a two-way flow of people between town and country, according to individual desires and national need? It all needs to be thought about. It is clear, though, that the New Agrarianism is necessary. The problems need to be addressed, in depth, asap.

5: How can we rescue science?

The world needs scientists. We need their rigor—their ability to find out what really is the case, and to suggest ways round what may seem intractable problems. We need, too, their moral fervor. Some of the very best people I know are scientists who combine personal unselfishness with truly impressive intellect, imagination, and a capacity for hard work that puts me to shame. But the ones I know who are truly touched with greatness are very often struggling to stay afloat in an academic atmosphere that is increasingly unhelpful, typically on low salaries and often with little or no security. The scientists who are doing well, materially, with big houses and cars and kudos, most often work for or are financed by biotech companies, convinced that the future of the world lies with GMOs, or just not giving a damn. They have no sense of the overall context, no idea of what traditional farming is, no conception of its strengths, but see traditional economies and ways of life as virgin territory, ripe for takeover.

If we truly want a world that is good for living in we have to reverse the trend of the past 35 years and begin again to reward scientists for working on behalf of humanity as a whole, rather than for the companies that make them personally rich.

6: How can we change minds?

I am not sure that minds need to be changed as much as is sometimes supposed. To be sure, many people have bought into the economy and politics of the past 35 years—they do apparently believe that their own material wealth is increasing, for iPods and Audis can be very seductive, and take it to be self-evident either that this is all that matters, or that what is true of them must also be true of everybody else as well, or that if people are falling by the wayside then it must be their own fault. But many others clearly are under no such illusions. Many are desperately frustrated, knowing that they really should not be working 60 hours a week just to put a roof over their heads and educate their children—though still unable to eat well—even though the world seems to be dripping with wealth. Even if they are doing well, many people clearly do care what happens to others and feel that their own lives are flawed, since their agreeable lifestyle is so obviously rooted in injustice. So it isn't really a question of changing minds; more of identifying the people who already know the kinds of things that need to be done, and bringing those people together, and creating a political and economic environment in which their inherent good sense and kindness can come to the fore.

These people are politically weak precisely because they do not go out of their way to seek power. Still, they are in the majority.

A College for Enlightened Agriculture seems just what is needed to address all these issues, and many others as they arise, and to go on addressing them. Together, the College for Enlightened Agriculture and the Worldwide Food Club could ensure that we, humanity, could at last begin to reclaim our own world and to put it back on an even keel. It may be too late. But it must be worth a try.

POSTSCRIPT

I t is hard to finish books.

The ink was no sooner dry on this one, metaphorically speaking, than I attended a conference at a very fine farm near my home that is both commercial and experimental: the farmers pioneer new approaches to husbandry, with special regard for animal welfare, but expect to make a living at the same time.

The chairman was a very distinguished biologist, and as might be expected of a senior professor and a knight of the realm, he spoke wisely. Indeed he reminded us all that however good and desirable their husbandry may be, farmers are dead in the water unless they can perform economically. There is no point, he said, in producing the most beautiful pork the world has seen, from pigs who have spent their lives in porcine paradise, if it costs $200 a kilo. He is perfectly right, of course. We all nodded sagely.

Yet there is a huge caveat.

At present, farmers the world over are being asked—obliged, coerced, forced with the full might of the law—to conform to an economic system that demonstrably is disastrous for farming in particular and for humanity and the world in general. The entire world economy is nothing more nor less than a contrived, largely ritualistic battle between corporates whose only brief, indeed whose *raison d'être*, is to generate as much disposable wealth as possible in the shortest possible time so as to attract the maximum number of shareholders; and of banks and various money-changers, whose task it is to shift the wealth around, once it has been generated. The world's most powerful governments, led by the neoconservatives of the United States but of course including those of Great Britain and of Russia, conform to this system absolutely. The present British government does not "govern" in any worthwhile sense. It merely provides the clearest possible arena for corporates to operate in; and its standard solution to all the problems of all countries, including those of much-beleaguered Africa, is to encourage corporates to set up shop within their boundaries.

Some of the CEOs of those corporates, and some members of some governments, truly believe that the choreographed battle of corporates and banks will somehow produce a world that is secure—one that can last another 10,000 years or a million years—and is agreeable to live in. Others, very obviously, do not give a damn. Their ambition is to maximize their own power in the short term. The long-term future seems to be beyond their imagining. But then again, some of them apparently believe in Armageddon. They seem to feel that the end is nigh whatever we do, so they might as well enjoy themselves while they can.

But whatever the individual psychology of the people with influence, the powers-that-be, it seems obvious even on grounds of simplest common sense that the economic system that they champion is most unlikely to produce a world that is secure and agreeable. It is not designed to do so, and indeed in many respects it seems intended to do the precise opposite—to sacrifice everything to the here-and-now advantage of a minority. The empirical evidence all around us suggests that what common sense declares is likely to be the case, is indeed the case. The present economic-political system is producing a world that for a very fair slice of humanity is already desperate, and cannot possibly be sustained. The only question seems to be whether enough of what's worthwhile can survive to make any kind of rescue possible, when the present madness has run its course.

So of course the professor is right, and all those other intellectuals and sages who tell us that we have to gear what we do to economic reality. But what he and the rest of us need to ask as a matter of urgency is whether the present economy, to which we are all being obliged to adjust, really is serving our interests, and those of planet Earth as a whole. Very obviously it is not. In all fields, we are being asked to sacrifice much that is good—indeed, virtually everything that humanity has evolved this past few tens of thousands of years, our crafts and traditions and ways of life—in favor of a power structure that must be seen either to be mistaken to the point of absurdity, or cynical beyond measure.

This generalization applies to all modern life. But it applies to agriculture in particular. Agriculture really is different. The present economic system may well produce the best possible battleships or computers or motor cars. If so, so be it. But it cannot possibly produce the best agriculture. The world needs farming that is specifically designed to feed people, and to provide agreeable ways of life, and to look after the environment— what I am calling Enlightened Agriculture. The demands of Enlightened Agriculture, as I have tried to show in this book, are absolutely at odds with those of the present economic system; as diametrically opposed as can be conceived. If agriculture fails, we will all be dead in weeks. To compromise agriculture, indeed to wreck it, in favor of an economic system that clearly is not intended to serve the real needs of humankind, is the most extraordinary nonsense. Yet the nonsense prevails.

To be sure, then, the professor is right in principle. Agriculture must be geared to economic reality. But the economy in turn must be geared to biological and social reality. Clearly, at present, this is not the case. The future, then—the future of all humanity—does not lie with farmers who can adjust what they do to the present economic norms, but to people at large, including farmers, who can create a new food supply chain, distinct from the present-day economy devised by the powers-that-be. The long-term task is to devise agriculture that really can produce good food forever (Enlightened Agriculture); to do so within a society that is agreeable to live in and is built around that agriculture (the New Agrarianism—based on science-assisted craft); and to devise an economic structure that is suited to the real needs of humanity and of the whole world (the New Capitalism).

The powers-that-be really have lost the plot. Simply to follow their lead, to accept the *status quo*, and to contrive to plug ourselves into the system they have evolved, is suicide.

Pari Publishing is an independent publishing company, based in a medieval Italian village. Our books appeal to a broad readership and focus on innovative ideas and approaches from new and established authors who are experts in their fields. We publish books in the areas of science, society, psychology, and the arts.

Our books are available at all good bookstores or online at **www.paripublishing.com**

If you would like to add your name to our email list to receive information about our forthcoming titles and our online newsletter please contact us at **newsletter@paripublishing.com**

Visit us at **www.paripublishing.com**

Pari Publishing Sas
Via Tozzi, 7
58040 Pari (GR)
Italy

Email: info@paripublishing.com